Disclaimer

The publisher of this book is by no way associated with the National Institute of Standards and Technology (NIST). The NIST did not publish this book. It was published by 50 page publications under the public domain license.

50 Page Publications.

Book Title: Test Methods for Quantifying the Propensity of Cigarettes to Ignite Soft Furnishings. Volume 2. [Also included is: Cigarette Extinction Test Method]

Book Author: Thomas J. Ohlemiller; K M. Villa; E Braun; K Eberhardt; Richard H. Harris; James R. Lawson;

Book Abstract: Research funded under the Fire Safe Cigarette Act of 1990 (P.L. 101-352) has led to the development of two test methods for measuring the ignition propensity of cigarettes. The Mock-Up Ignition Test Method uses substrated physically similar to upholstered furniture and mattresses: a layer of fabric over padding. The measure of cigarette performance is ignition or non-ignition of the substrate. The Cigarette Extinction Test Method replaces the fabric/padding assembly with multiple layers of common filter paper. The measure of performance is full-length burning or self-extinguishment of the cigarette. Routine measurement of the relative ignition propensity of cigarettes is feasible using either of the two methods. Improved cigarette performance under both methods has been linked with reduced real-world ignition behavior; and it is reasonable to assume that this, in turn, implies a significant real-world benefit. Both methods have been subjected to interlaboratory study. The resulting reproducibilities were comparable to each other and comparable to those in other fire test methods currently being used to regulate materials which may be involved in unwanted fires. Using the two methods, some current commercial cigarettes are shown to have reduced ignition propensities relative to the current best-selling cigarettes. [*] This is one of six volumes in the Final Report, Fire Safe Cigarette Act of 1990. VOLUME 1. Overview: Practicability of Developing a Performance Standard to Reduce Cigarette Ignition Propensity by Jones-Smith, J., et al. VOLUME 3. Modeling the Ignition of Soft Furnishings by a Cigarette by Mitler, H. E., et al. VOLUME 4. Cigarette Fire Incident Study by Harwood, B., et al. VOLUME 5. Toxicity Testing Plan by Lee, B. C., et al. VOLUME 6. Societal Costs of Cigarette Fires by Ray, D. R., et al.

Citation: NIST SP - 851

Keyword: cigarettes; test methods; ignition; upholstered furniture; statistical analysis; self-extinguishment

Test Methods for Quantifying the Propensity of Cigarettes to Ignite Soft Furnishings

2

U.S. Department of Commerce

Technology Administration

National Institute of Standards and Technology

August 1993

NIST Special Publication 851

NIST

Fire Safe Cigarette Act of 1990

Under the Cigarette Safety Act of 1984 (P.L. 98-567), the Technical Study Group on Cigarette and Little Cigar Fire Safety (TSG) found that it is technically feasible and may be commercially feasible to develop a cigarette that will have a significantly reduced propensity to ignite furniture and mattresses. Furthermore, they found that the overall impact of such a cigarette on other aspects of the United States society and economy may be minimal.

Recognizing that cigarette-ignited fires continue to be the leading cause of fire deaths in the United States, the Fire Safe Cigarette Act of 1990 (P.L. 101-352) was passed by the 101st Congress and signed into law on August 10, 1990. The Act deemed it appropriate for the U.S. Consumer Product Safety Commission to complete the research recommended by the TSG and provide, by August 10, 1993, an assessment of the practicality of a cigarette fire safety performance standard.

Three particular tasks were assigned to the National Institute of Standards and Technology's Building and Fire Research Laboratory:

- develop a standard test method to determine cigarette ignition propensity,

- compile performance data for cigarettes using the standard test method, and

- conduct laboratory studies on and computer modeling of ignition physics to develop valid, user-friendly predictive capability.

Three tasks were assigned to the Consumer Product Safety Commission:

- design and implement a study to collect baseline and follow-up data about the characteristics of cigarettes, products ignited, and smokers involved in fires,

- develop information on societal costs of cigarette-ignited fires, and

- in consultation with the Secretary of Health and Human Services, develop information on changes in the toxicity of smoke and resultant health effects from cigarette prototypes.

The Act also established a Technical Advisory Group to advise and work with the two agencies.

This report is one of six describing the research performed and the results obtained. Copies of these reports may be obtained from the **U.S. Consumer Product Safety Commission, Washington, DC 20207.**

2

Test Methods for Quantifying the Propensity of Cigarettes to Ignite Soft Furnishings

Thomas J. Ohlemiller [a]
Kay M. Villa [a,c]
Emil Braun [a]
Keith R. Eberhardt [b]
Richard H. Harris, Jr. [a]
J. Randall Lawson [a]
and Richard G. Gann [a]

[a] Building and Fire Research Laboratory
[b] Computing and Applied Mathematics Laboratory
 National Institute of Standards and Technology
 Gaithersburg, Maryland 20899
[c] Current Affiliation: American Textile
 Manufacturers Institute

U.S. Department of Commerce
Ronald H. Brown, Secretary
Technology Administration
Mary L. Good, Under Secretary for Technology
National Institute of Standards and Technology
Arati Prabakhar, Director

NIST Special Publication 851
August 1993

National Institute of Standards and
Technology
Special Publication 851
Natl. Inst. Stand. Technol.
Spec. Pub. 851
166 Pages (Aug. 1993)
CODEN: NSPUE2

U.S. Government Printing Office
Washington: 1993

For sale by the Superintendent of
Documents
U.S. Government Printing Office
Washington, DC 20402-9325

TABLE OF CONTENTS

Page

List of Tables v

List of Figures vii

EXECUTIVE SUMMARY ix

I. INTRODUCTION 1

    A. Perspective on the Current Projects 1

    B. General Considerations Regarding Test Methods 3
       1. Applications 3
       2. Output 3
       3. Figure of Merit 4
       4. Validity 5
       5. Long-Term Utility 6

II. TEST METHODS DEVELOPED IN THE PRESENT STUDY 7

    A. Cigarettes Used in the Present Study 7
       1. Series 100 Experimental Cigarettes 7
       2. Series 500 Experimental Cigarettes 11

    B. Mock-Up Ignition Test Method 13
       1. Previous Use of Mock-ups 13
       2. Fabric Considerations for a Mock-Up Test Method 14
       3. Other Mock-up Materials 26
       4. Mock-up Configuration 28
       5. Enclosure Design; Air Flow Considerations 30
       6. Test Variables 33
       7. General Description of the Mock-Up Ignition Test Method .. 36
       8. Interlaboratory Study of the Mock-Up Method 37

    C. Cigarette Extinction Test Method 70
       1. Prior Alternative Methods 70
       2. Approaches Examined in This Study 71
       3. Standard Materials 83
       4. Enclosure Design 85
       5. General Description of the Test Method 85
       6. Interlaboratory Study of the Test Method 85

III.	CONSIDERATIONS REGARDING THE USE OF THE TWO TEST METHODS		99
	A.	Mock-Up Ignition Test Method	99
	B.	Cigarette Extinction Test Method	102
	C.	Allowable Material Variability	102
	D.	Standardization of Test Methods	104
	E.	Effectiveness of the Methods	104
IV.	TESTING OF COMMERCIAL CIGARETTES		106
	A.	Introduction	106
	B.	Rationale for Commercial Cigarette Choices	106
	C.	Test Procedures	107
	D.	Data and Analysis	107
V.	CONCLUSIONS AND RECOMMENDATIONS		110
VI.	ACKNOWLEDGEMENTS		111
VII.	REFERENCES		112

APPENDIX A:	Reevaluation of Experimental Cigarettes Used in the Cigarette Safety Act of 1984	A-1
APPENDIX B:	Mock-Up Ignition Test Method Procedure	B-1
APPENDIX C:	Estimate of Oxygen Supply Paths to a Cigarette Coal atop a Flat Upholstery Substrate	C-1
APPENDIX D:	Metal Ion Content of Fabrics: Test Method and Results	D-1
APPENDIX E:	Cigarette Extinction Test Method Procedure	E-1
APPENDIX F:	Representative Thermogravimetric Data for Test Method Materials	F-1

LIST OF TABLES

		Page
1.	Description of Series 100 Experimental Cigarettes	9
2.	Ignition Propensity of Series 100 Experimental Cigarettes	10
3.	NIST Comparison of Series 100 and 500 Cigarette Weights	11
4.	Cigarette Industry Comparison of Series 100 and 500 Cigarette Weights	12
5.	Comparison of Ignition Propensities for Series 100 and 500 Cigarettes	13
6.	Specified Nominal Properties of Fabrics	17
7.	Cation Content of Fabrics Used in the Preliminary and Main Interlaboratory Studies	21
8.	Potassium Content of West Point Pepperell Cotton Ducks Over a Four Month Period	22
9.	Sensitivity of Ignition Susceptibility to K^+ Content in Fabrics	23
10.	Ignition Susceptibility of Different #8 Cotton Duck Fabric Samples	23
11.	Measured Areal Densities of Fabrics Used In Interlaboratory Study	25
12.	Measured Air Permeability of Fabrics Used in Interlaboratory Study and in TSG Study	25
13.	Sensitivity of Ignition Susceptibility to Foam Properties	26
14.	Measured Air Permeability of Polyurethane Foam By ASTM D 3574	27
15.	Properties of Polyethylene Films Used in Conjunction with Duck #4	27
16.	Percent Ignitions on Various Substrates for Selected Cigarettes	29
17.	Effect of Air Flow Disturbance on Cigarette Ignition Propensity	32
18.	Estimated Sensitivity of Mock-up Test Outcome to Test Variables	35
19.	Description of Interlaboratory Study Cigarettes	40
20.	Variables in Analysis of Preliminary Interlaboratory Study	44

21.	Summary of Test Results for Preliminary Interlaboratory Study	46
22.	Summary of Test Results for Main Interlaboratory Study, Mock-Up Ignition Method	55
23.	Observed Repeatability and Reproducibility Standard Deviations for Mock-Up Ignition Method, Main Interlaboratory Study	65
24.	Mock-Up Ignition Method: Calculated Repeatability and Reproducibility Limits for Various Assumed Numbers of Replications and Ignition Propensities	69
25.	Large Glass Bead Non-Reactive Substrate Test Results for Selected Cigarettes and Air Speeds	75
26.	Re-test of Selected Cigarettes on Large Glass Bead Substrate	75
27.	Variability of Filter Paper Areal Density and Thickness	83
28.	Summary of Test Results for Interlaboratory Study of Cigarette Extinction Method	87
29.	Observed Repeatability and Reproducibility Standard Deviations for Cigarette Extinction Method Interlaboratory Study	96
30.	Cigarette Extinction Method: Calculated Repeatability and Reproducibility Limits for Various Assumed Numbers of Replications and Full-Length Burn Proportions	97
31.	95% Lower Confidence Bounds for the Long-Run Ignition Probability	101
32.	Results of Commercial Cigarette Testing	108
33.	Percent Ignitions or Full Length Burns on Test Method Substrates	109
34.	Averaged Smoke Component Yields from Commercial Cigarettes	109
A-1.	Selection of Cigarettes for Reevaluation Study	A-3
A-2.	Reevaluation of Eight of the Thirty-Two Experimental Cigarettes	A-4
C-1.	Calculated Mass Flux of Oxygen From Foam to Coal	C-3
D-1.	Cation Content of #4 Cotton Duck	D-2
D-2.	Cation Content of #6 Cotton Duck	D-3
D-3.	Cation Content of #10 Cotton Duck	D-4

LIST OF FIGURES

		Page
1.	Photograph of a Test Chamber Containing a Mock-Up Assembly and a Cigarette	38
2.	Comparison of Ignition Rates for the Preliminary Interlaboratory Study of the Mock-up Ignition Test Method	48
3.	Environmental Conditions Reported by the Laboratories Participating in the Preliminary Interlaboratory Study of the Mock-up Ignition Test Method	49
4.	Comparison of Ignition Rates for the Main Interlaboratory Study of the Mock-up Ignition Test Method	60
5.	Environmental Conditions Reported by the Laboratories Participating in the Main Interlaboratory Study of the Mock-up Ignition Test Method	62
6.	Plot Showing Empirical Relation of Reproducibility Variance to Repeatability Variance for the Mock-Up Ignition Test Method	68
7.	Free Burning Rate of Various Cigarettes Suspended in Quiescent Air as a Function of the Fraction of the TSG Mock-Up Failures	72
8.	Schematic Representation of the Test Assembly for the Glass Bead/Rod Substrate Tests	74
9.	Drawing of the Cigarette Thermal Transfer Test Assembly	77
10.	Typical Average Temperature-Time Trace for a Cigarette Burning on a Glass Fiber Filter Paper in the Cigarette Thermal Transfer Test Assembly	78
11.	Estimated Energy Transferred to a Substrate from a Smoldering Cigarette Burning in the Thermal Transfer Apparatus as a Function of the Fraction of TSG Mock-Up Failures	79
12.	Smoldering Rates of Three Experimental Cigarettes as a Function of the Number of Filter Papers Making up the Substrate Assembly	81
13.	Number of Filter Papers Causing Extinguishment of the Cigarette as a Function of the TSG Failure Fraction	82
14.	Photograph of a Test Chamber Containing a Mock-Up Assembly and a Cigarette	84
15.	Comparison of Full Burn Rates for the Interlaboratory Study of the Cigarette Extinction Test Method	93

16.	Environmental Conditions Reported by the Laboratories Participating in the Interlaboratory Study of the Cigarette Extinction Test Method	94
17.	Plot Showing Empirical Relation of Reproducibility Variance to Repeatability Variance for the Cigarette Extinction Test Method	98
B-1.	Schematic of Test Chamber Components	B-7
B-2.	Schematic of Vacuum Draw Apparatus	B-8
B-3.	Location of Cigarette on Mock-Up Method Substrate Assembly	B-9
E-1.	Details of the Filter Paper Holder Support Structure	E-7
E-2.	Brass Holddown Ring and Cigarette Motion Restrainers	E-8
F-1.	TG Behavior of Two Samples of Cotton Duck #4	F-2
F-2.	TG Behavior of Two Samples of Cotton Duck #6	F-3
F-3.	TG Behavior of Two Samples of Cotton Duck #10	F-4
F-4.	Derivative TG Data: Two Polyurethane Foam Samples from Top and Bottom of Original Bun	F-5
F-5.	TG Data for Two Samples of Polyethylene Film	F-6
F-6.	TG Data for Two Samples of Whatman Filter Paper	F-7

TEST METHODS FOR QUANTIFYING THE PROPENSITY OF CIGARETTES TO IGNITE SOFT FURNISHINGS

Thomas J. Ohlemiller[a], Kay M. Villa[a,b], Emil Braun[a], Keith R. Eberhardt[c],
Richard H. Harris, Jr.[a], J. Randall Lawson[a], and Richard G. Gann[a]

[a] Building and Fire Research Laboratory
[c] Computing and Applied Mathematics Laboratory
National Institute of Standards and Technology
Gaithersburg, MD 20899

[b] Current Affiliation: American Textile Manufacturers Institute

EXECUTIVE SUMMARY

Cigarette ignition of soft furnishings (upholstered furniture and mattresses) continues to be the leading cause of fire deaths in the United States. In 1990, the nation experienced 1220 lost lives, 3358 serious civilian injuries, and $400 million in direct property loss from 44,000 cigarette-initiated fires in structures. This publication describes the research performed and the results obtained in responding to two tasks under the Fire Safe Cigarette Act of 1990 (P.L. 101-352):

"(1) develop a standard test method to determine cigarette ignition propensity, and

(2) compile performance data for cigarettes using the standard test method developed under paragraph (1)"

as part of an assessment of the practicability of developing a performance standard to reduce cigarette ignition propensity. This research builds on previous studies directed by the Technical Study Group (TSG) under the Cigarette Safety Act of 1984 (P.L. 98-567) and related work performed by the cigarette industry.

The principal content of the report is documentation of the selection, development and final form of two test methods for cigarette ignition propensity. They are intended to fulfill two potential roles: (a) the basis for a possible performance standard, and (b) assistance to the cigarette industry in meeting the goals of any such regulation and in quality assurance testing. Both methods have valid links (comparable to many current fire test methods) to many real-world fire scenarios of concern. Both incorporate most of the relevant physics and chemistry of such ignitions, while replicating the real-world hazard to differing extents. They are both performance tests, as contrasted with product design specifications. Both tests offer the use of a graded measure of performance, where acceptable levels can be set by the regulator. The research and this report do not address specific regulatory criteria.

The *Mock-Up Ignition Test Method* uses three types of simulated upholstery cushions, each with a different cigarette ignition susceptibility. Each 20 cm x 20 cm assembly (substrate) consists of a top layer of one of three weights of cotton duck fabric (#4, #6, and #10, in increasing order of ignition susceptibility); a 5 cm thick piece of a polyurethane foam; and, in the least susceptible substrate, a thin layer of thermoplastic film in between. Tests are conducted in a plastic enclosure to eliminate variability due to laboratory air currents. A test begins by placing a lit cigarette on the mock-up. The performance measure is whether or not the mock-up is ignited (char propagation over 10 mm from the burning tobacco column). Either self-extinction of the cigarette or the cigarette burning its entire length without igniting the mock-up assembly are counted as non-ignitions. A complete test series consists of 48 replicates of each cigarette on each substrate.

The *Cigarette Extinction Test Method* replaces the more complex substrate of the Mock-Up Ignition Test Method with standard cellulosic filter paper. Otherwise, the test procedure is similar. The three substrates used consist of 3, 10, or 15 layers of the paper. The test determines whether a selected substrate absorbs enough heat from the cigarette coal to extinguish the cigarette. Performance with this method was roughly correlated to prior direct measures of cigarette ignition propensity. Here, increased reproducibility of test materials is gained at the cost of direct simulation of the real-world fire scenario. Sixteen replicates of each cigarette are performed on each substrate.

Only flat substrates were selected, although many real-world ignitions are expected to occur in furniture crevices. The TSG studies showed a higher fraction of crevice ignitions for a cigarette of high ignition propensity, but no consistent difference in ignition susceptibility between the two configurations for cigarettes of moderate-to-low ignition propensity. Potential variability of contact between the cigarette coal and the surfaces of the crevice substrates introduces an operator dependence that is undesirable.

All testing is performed without externally-imposed air flow. This is operationally the simplest approach and is highly relevant. In the real world, the orientation of any flow relative to the cigarette coal is unknown but probably random. Many ignitions may occur deep in a crevice, and the air flow there is likely to be very small. While cigarette industry studies showed some cigarettes undergoing substantial changes in rankings of ignition performance under varying air flows, greater flow differences between mock-up and chair tests in the TSG studies did not preclude a good correlation between these two types of tests. The existence of this correlation strongly implies that there will be a real-world benefit in moving toward cigarettes which perform well in the two test methods developed here. Should further information on real-world ignitions indicate a significant fraction due to low ignition propensity cigarettes in external air flow conditions *at the ignition location*, it may be appropriate to supplement the results of the current methods with those of tests conducted in the presence of a comparable flow.

The two test methods were developed using experimental cigarettes manufactured by the cigarette industry for this purpose. The cigarettes varied widely in performance, from some having ignition propensities comparable to current commercial cigarettes to others that rarely or never ignited any of the test substrates in both this and cigarette industry studies.

The two methods were shown to be of useful reproducibility in a nine-laboratory study. The study involved cigarette industry, state and federal agency, and private testing laboratories, and conformed to ASTM guidelines. Five of the available experimental cigarettes were tested, based on their expected ignition performance.

The *repeatability* (a measure of variability within a laboratory) decreases as the square root of the number of replicates. Thus, for production quality assurance testing, a fine degree of resolution is possible. By contrast, the *reproducibility* (a measure of variability between laboratories) approaches a non-zero limit for a large number of replicates. Typically, for both of these test methods, the ASTM reproducibility limit of the percentage of ignitions or the percentage of cigarettes burning their full lengths on a given substrate was ca. 40 percent. This value defines the limit of resolution for use in any future regulations.

The study showed that the lab-to-lab variability of results was comparable to that for other fire test methods currently being used to regulate materials which may be involved in unwanted fires. The results were generally insensitive to the date and time of day of testing, the particular test enclosure used, and the operator skill level. All labs conformed sufficiently to the temperature and humidity criteria for the conditioning and test rooms that this was not an important factor in the results. The three substrates in each method were all statistically distinct from each other, as were the five cigarette types.

Since the results show that the methods can effectively differentiate the ignition propensities of various cigarettes with acceptable precision, specifications for the test materials were developed. All four types of materials were deemed likely to be available, with long-term consistency, in the foreseeable future. For the fabrics, the areal density and potassium ion content were determined to be the major parameters affecting ignition susceptibility. Analysis of within-lot samples, lot-to-lot samples, and samples from two manufacturers showed that the normal production variations were within the acceptable limits demonstrated in the interlaboratory study. There is a long history of a large demand for cotton duck fabrics for both commercial use and military procurement. The polyurethane foam is representative of foam products used in the residential furniture market. Experiments showed that the effect of expected foam property variations (within nominally similar formulations) is minimal. Differences between brands of purportedly the same polyethylene film resulted in a significant change in test method results. However, specification of the areal density should ensure use of a proper material. The filter paper is a long-time, high-purity standard material for numerous chemical methods. Variations in the areal density, thickness and thermal conductivity are minimal. It was estimated that "fresh" substrate materials did not age substantially over about 6 months or longer.

There are data to "calibrate" the methods at the high and low ends of the ignition propensity scale. The commercial cigarette data in the TSG studies establish an indication of performance for the cigarettes associated with then-current fire losses. In the two new test methods, this performance is seen as a large number of ignitions on the #4 cotton duck or full-length burning on the 15-layer paper substrate. This establishes the test results for the high ignition propensity end of the scale. The TSG work, the current research, and cigarette industry studies demonstrate that there are experimental cigarettes that never or rarely ignited a variety of substrates. In the two new test methods, this behavior is observed as few ignitions on the #10 cotton duck or few full-length burns on 3 layers of filter paper. In between these extremes, one would like to expect a reduced number of fires as fewer ignitions are measured in the laboratory. The TSG correlation of mock-up results with chair tests indicates that such results can be expected to be indicative of performance for significant portions of the real-world furniture population, at least for coarse changes in test performance. If considering small increments, however, one must keep in mind the accuracy limits of the methods as discussed above.

For a product standard, there is a preference at present for using the Mock-Up Ignition Test Method, because it is capable of better distinction among cigarettes of high ignition propensity. However, routine measurement of the relative ignition propensity of cigarettes is feasible using either of the two methods. The mock-up ignition method requires about 3 staff days to perform the 144 tests; the cigarette extinction method, with its simpler substrates and 48 tests, about 1 staff day. A rationale has been developed to reduce the number of tests for cigarettes of expected very high or very low ignition propensity.

It is common practice, upon development of a fire test method for professional use, to proceed with its adoption as a voluntary consensus standard in either the ASTM or the National Fire Protection Association (NFPA). This report contains sufficient documentation of the two test methods and interlaboratory evaluations of each. Thus, all necessary materials for initiating the standardization process are now available.

Twenty current commercial cigarettes were tested using the two methods. Fourteen of these were the best-selling packings, comprising nearly 40 percent of total sales in 1990. These cigarettes did not vary widely in their physical characteristics. They showed consistent ignitions on all substrates using the Mock-Up Ignition Method and consistently burned their full length on all substrates tested in the Cigarette Extinction Method.

Also tested were six other packings, each having one or two physical parameters (*e.g.*, low circumference, paper porosity, tobacco density) which deviate from the best-sellers in a direction which prior research would suggest as likely to lower ignition propensity. All six of these packings showed reduced ignition propensity in the Mock-Up Ignition Test Method. Four of these packings rarely ignited the most difficult-to-ignite substrate; the other two ignited it in 40-70% of the tests. Three of the four packings showed reduced ignition propensity on the middle substrate as well. While the Cigarette Extinction Test Method is less sensitive to changes in ignition propensity, three of the packings showed markedly fewer full-length burns. All these differentiations are outside the variability of the test methods. The average values of tar, nicotine, and carbon monoxide yields for these six packings were no larger than the averages for the 14 best-selling cigarettes.

TEST METHODS FOR QUANTIFYING THE PROPENSITY OF CIGARETTES TO IGNITE SOFT FURNISHINGS

Thomas J. Ohlemiller, Kay M. Villa, Emil Braun, Keith R. Eberhardt,
Richard H. Harris, Jr., J. Randall Lawson, and Richard G. Gann

ABSTRACT

Research funded under the Fire Safe Cigarette Act of 1990 (P.L. 101-352) has led to the development of two test methods for measuring the ignition propensity of cigarettes. The Mock-Up Ignition Test Method uses substrates physically similar to upholstered furniture and mattresses: a layer of fabric over padding. The measure of cigarette performance is ignition or non-ignition of the substrate. The Cigarette Extinction Test Method replaces the fabric/padding assembly with multiple layers of common filter paper. The measure of performance is full-length burning or self-extinguishment of the cigarette. Routine measurement of the relative ignition propensity of cigarettes is feasible using either of the two methods. Improved cigarette performance under both methods has been linked with reduced real-world ignition behavior; and it is reasonable to assume that this, in turn, implies a significant real-world benefit. Both methods have been subjected to interlaboratory study. The resulting reproducibilities were comparable to each other and comparable to those in other fire test methods currently being used to regulate materials which may be involved in unwanted fires. Using the two methods, some current commercial cigarettes are shown to have reduced ignition propensities relative to the current best-selling cigarettes.

Key words: Fire, cigarettes, cigarette test method, ignition, upholstered furniture, statistical analysis

I. INTRODUCTION

A. Perspective on the Current Projects

Cigarette ignition of soft furnishings (upholstered furniture and mattresses) continues to be the leading cause of fire deaths in the United States.[1] In 1990, the nation experienced 1220 lost lives, 3358 serious civilian injuries, and $400 million in direct property loss from 44,000 cigarette-initiated fires in structures. These figures continue a slow downward trend (except in property loss, which is increasing) with cause(s) suggested but not established.

As a means to accelerate reducing these losses, the Cigarette Safety Act of 1984 (Public Law 98-567) created a Technical Study Group on Cigarette and Little Cigar Fire Safety (hereafter, TSG) and directed it to:

> "undertake such studies and other activities as it considers necessary and appropriate to determine the technical and commercial feasibility, economic impact, and other consequences of developing cigarettes and little cigars that will have a minimum propensity to ignite upholstered furniture or mattresses."

In its final report [2], the TSG concluded that:

> "It is technically feasible and may be commercially feasible to develop cigarettes that will have a significantly reduced propensity to ignite upholstered furniture or mattresses."

However, in assessing the commercial feasibility, the TSG membership also noted that:

> "A valid and reliable test method is needed to measure the reduced ignition propensity of improved cigarettes."

> "... the current mockup method is usable for research measurements of the relative ignition propensity of cigarettes. However, because of the lot-to-lot variability of the fabrics and paddings used, this method should not be used as the standard test method."

> "None of the several alternative candidate test methods for measuring the cigarette ignition propensity of soft furnishings was usable in its current state of development."

These statements reaffirm what has been found for many products: desired performance must be *measurable*. This quality allows a specifier to declare what is expected of the product, the manufacturer to produce a desired commodity, and the vendor to demonstrate compliance with the specifier's demands. A standardized performance measurement or test method makes this possible. It then becomes the role of society to determine the level of performance it desires and how much it is willing to pay. It is noteworthy that several state legislatures have delayed mandating less fire-prone cigarettes for lack of a quantitative test method.

Recognizing this as a key link to reducing fire losses, the Congress enacted the Fire Safe Cigarette Act of 1990 (P.L. 101-352), noting that:

> "It is appropriate for the Congress to require by law the completion of the research described in the final report of the Technical Study Group on Cigarette and Little Cigar Fire Safety and an assessment of the practicability of developing a performance standard to reduce cigarette ignition propensity, and

> it is appropriate for the Consumer Product Safety Commission to utilize its expertise to complete the recommendations for further work and report to Congress in a timely fashion."

Accordingly, the Act directed that the National Institute of Standards and Technology's Center for Fire Research [now the Building and Fire Research Laboratory (BFRL)], at the request of the Consumer Product Safety Commission (CPSC):

> "(1) develop a standard test method to determine cigarette ignition propensity,

> (2) compile performance data for cigarettes using the standard test method developed under paragraph (1), and

> (3) conduct laboratory studies on and computer modeling of ignition physics to develop valid, user-friendly predictive capability."

This publication describes the research performed and the results obtained in responding to the first two tasks. NIST has developed two test methods with sound links to the real-world fire scenarios of concern. These methods were shown to be of useful reproducibility in a nine-laboratory evaluation. The methods were then used to evaluate a sampling of the most popular current commercial cigarettes, as well as some whose physical properties suggest they might show reduced ignition propensity. The completion of the third task is described in a companion report.

B. General Considerations Regarding Test Methods

There are several ways of describing test methods and the features that are necessary for their use in professional fire safety practice. The following sections describe these in the context of the current program.

1. Applications

The test methods developed here are intended to fulfill two potential roles. The first role is to serve as a *practical* basis for a possible performance standard. As stated earlier, a regulation presupposes the existence of a practical test method. It is not feasible to make cigarette ignition propensity assessments on a recurring basis by testing each cigarette type on all soft furnishings in the commercial marketplace because:

- The upholstered furniture market is extremely diverse and not well-defined in terms of the materials used and their market shares.

- Usage may cause soft furnishings to respond differently to contact with lighted cigarettes, perhaps as a consequence of such factors as fabric wear, the use of cleaning fluids, or alkali metal accumulation.

- The resources needed for such an approach would be prohibitive.

A second role for an ignition propensity test method is to assist the cigarette industry in meeting the goals of any such regulation. This has two potential applications:

- Guidance in product development, in which the test results are used to indicate progress toward more desirable ignition behavior; and

- Quality assurance on the production line, in which sample cigarettes taken at intervals are checked to ensure they meet the regulatory requirement.

2. Output

The output of a test method can be continuous, discrete, or pass/fail. In this order, the methods produce a decreasing amount of information to the regulator, product developer, and performance monitor. An example of the first is automobile gas mileage testing, where any value of miles per gallon may result from the dynamometer test measurements. The regulator then selects a value from the continuum as the acceptable product characterization. In test methods with discrete output, only a fixed number of results are possible. An example might be marksman ratings, which are based on the number of "hits" from a selected number of shots fired. Another example is tire traction ratings which place all results in a small number of categories. In each of these two examples, one obtains qualitative information about the performance of the product, relative to both the scale for evaluation and to other products. By contrast, a pass/fail test *only* provides an indicator of acceptability. For example, if you cannot read the eye chart correctly, you won't qualify for a driving license.

It is possible for a single type of test apparatus to be used in multiple modes. Consider the upholstered furniture mock-up experiments performed under the TSG program [3]. One could perform 10 ignition tests for a given cigarette on each of 3 mock-up constructions. A possible *continuous* output could be the mean time for ignition to occur, taking into account those tests that did not result in ignition. A *discrete* measure could be the number of tests that led to ignition. A *pass/fail* use might dictate that no cigarette burn longer than 1 minute on the mock-up.

As can be seen from the above examples, all of these types of methods are acceptable in everyday usage. However, it is preferable *but not mandatory* for product regulation that a test method provide a graded measure of performance. In this context it then becomes important to quantify the level of precision warranted by the measurements. This includes both the degree to which a single tester will reproduce the same result in multiple tests (repeatability) and the range of results that would be obtained when different testers perform the procedure (reproducibility). This will be discussed further in a later section.

3. Figure of Merit

Test methods may also be grouped by what it is that they measure. A *design* or *property* test measures a physical or chemical feature of the product. Thus, utilizing such a test method one might (improperly) extrapolate the results of the TSG study [2] and require that all cigarettes should be fabricated of tobacco below a prescribed packing density, be of less than a prescribed circumference, and be fabricated using paper of air permeability below a prescribed value. Alternatively, an index could be prescribed combining these factors. This kind of test presumes that the other descriptors of the product do not affect the desired performance. The result of a prescriptive regulation based on a property test is a (partial) description of the product.

By contrast, a *performance* test simulates the conditions of the (undesirable) outcome of the product's use. The TSG furniture mock-up testing is a convenient example. A regulation based on this kind of test would not directly dictate the physical nature of the cigarette. However, it might impose subtle limitations. For example, a 5 cm x 5 cm mock-up surface could not support a 15 cm long cigarette while exposing the fabric to the coal. Very long cigarettes would thus be discriminated against by the method, possibly restricting their introduction into the marketplace.

The degree to which a performance test replicates the potential hazard leads to further considerations. Ideally, the test should mimic the actual cigarette-initiated fire conditions as closely as possible. Since the critical elements of these conditions are simply the cigarette and its immediate environs, this would seem to be readily achievable. One need only abstract the region of the upholstered chair, sofa or mattress that influences the ignition process and incorporate it in the test, effectively achieving a full-scale simulation of the real-world hazard. In practice, these environs are not unchanging; they may vary appreciably with furniture design and materials, as well as with the chance aspects of cigarette contact. Thus one or more realistic examples are chosen, an approach embodied in the use of upholstery mock-ups. A test consists of placing a lit cigarette on some small-scale configuration of a cushion covered by an upholstery fabric and observing the consequences. If a smolder zone develops in the fabric and spreads continually away from the cigarette coal, the cigarette has failed the test. Successful tests such as this incorporate most of the relevant physics and chemistry, while not necessarily replicating the real world hazard exactly.

The other general orientation which a performance test method could take is to measure some aspect of the cigarette which has been shown to correlate with its tendency to ignite upholstered furniture. Such correlating features of the cigarette are not readily discerned. It is certainly useful to have some insights into the physics of the ignition process in order to pursue this approach. Ihrig *et al.* [4] examined a large number of upholstery fabrics and a small number of cigarettes. They inferred that only the total radiative heat output of a cigarette (joules/cig.) was a useful predictor of ignition propensity. Gann *et al.* [3] examined a wide variety of experimental cigarettes as part of a detailed study of the physics of the ignition process, but found no single *performance* parameter which gave a strong correlation with ignition propensity. The current study has been more successful in finding a performance measurement that correlates with ignition propensity, as described below.

4. Validity

The results of a performance test method must be linked to the real world; *e.g.*, for cigarette testing, there must be a direct correlation between the test method outcome and the real-world propensity to cause cigarette ignitions. As is often the case, this is a difficult matter here, because the actual condition (and thus ignition susceptibility) of in-use upholstered furniture cannot be well characterized.

For nearly all fire tests, the needed degree of reality is demonstrated by physical similarity between the test method and the real-world hazard and/or by use of the physical principles that determine fire initiation and growth. The principal basis for relating mock-up and full-scale behavior of furniture ignition by cigarettes is reported in reference 3. That study, while necessarily limited in the range of materials, chair configurations and number of test replicates, nevertheless established that:

- upholstery mock-ups can differentiate among cigarettes, and

- mock-up ignition behavior has shown a statistically-significant correlation with the behavior of full-scale chairs containing the same fabric and padding in the TSG study [2].

Evidence is presented in Section II.B that the substrates chosen for the Mock-Up Ignition Test Method are appropriate to represent actual upholstered furniture. The similarity of the two methods (with their different performance measures) in rating the performance of both experimental and commercial cigarettes (Section IV) lends credence to the validity of both.

5. Long-Term Utility

While the previous study revealed a set of mock-up material combinations capable of differentiating among cigarettes, it did not provide the necessary assurance of long-term test method reproducibility. The upholstery materials used there and, in fact, upholstery materials in general are not subject to any kind of quality control which bears on their ignitability by cigarettes. On the contrary, there have been indications that even a fabric such as California Standard cotton velvet [5], long used to assess the cigarette ignition resistance of flexible cushioning materials, has been inconsistent in its behavior [3]. One of the significant concerns of the present study has been to assess the factors which need to be controlled to assure long term consistency in mock-up response to cigarettes. The result of this work is a "Mock-Up Ignition Test Method" in which the substrates consist of cotton duck fabrics and a polyurethane foam. The details of the work which led to this method are presented in Section II.B.

To reduce further the dependence on substrate materials whose properties may be hard to assure on a long-term basis, a substantial effort has also been invested in developing a second test method. This "Cigarette Extinction Test Method" uses standard cellulosic filter paper as the sole material in contact with the tested cigarette. This method determines whether a selected number of layers of filter paper absorbs enough heat from the cigarette coal to extinguish the cigarette. Reproducibility of test materials is gained at the cost of evident physical similarity to the real-world fire scenario. Thus, a correlation with upholstery ignition measurements, as in Section IV, is necessary to establish the method's validity. A detailed description of this method is given in Section II.C.

II. TEST METHODS DEVELOPED IN THE PRESENT STUDY

A. Cigarettes Used in the Present Study

1. Series 100 Experimental Cigarettes

Series 100 refers to the series of cigarettes whose ignition propensities were measured in the previous study [3] (referred to throughout this report as the TSG study). They thus enable connecting the test methods developed in this study with the prior results. The 32 cigarettes were manufactured by the cigarette industry with then-current hardware at slower speeds. They varied systematically in five parameters at two levels, reported to be at the extremes of that equipment, with all other properties stated by the manufacturers to be identical, but not specified. The variable parameters and their values were:

- tobacco blend (Burley or flue-cured),

- tobacco expansion (nonexpanded, 60 cuts/inch; or expanded, 30 cuts/inch),

- cigarette circumference (nominally 21 or 25 mm),

- cigarette paper permeability (nominally 10 or 75 CORESTA units), and

- cigarette paper treatment (untreated or treated with approximately 0.8% sodium potassium citrate).

It should be noted that these experimental cigarettes may differ substantially from current commercial practice in having limiting values of some design parameters and in having no specification at all for other potentially pertinent parameters such as humectant or flavoring additive levels.

Table 1 gives the experimental cigarette designations with respect to the five parameters and the assigned cigarette numbers. Detailed information on the cigarettes can be found in the tables and appendices of Section 2 of reference 3, 3. Table 2 gives the ignition behavior of the TSG cigarettes summed over the four mock-up configurations used there.

The Series 100 cigarettes have been kept in cold storage (approximately -18 °C) since the end of the TSG study in 1987. Because approximately 4 years had elapsed between the two studies, changes in the cigarettes were possible. NIST thus undertook a reevaluation of the ignition propensity of the cigarettes on the same fabrics and padding materials used in the original TSG study. These had been stored in a nominally climate-controlled room ($\approx$ 21 °C, 30-60 % R.H.) since the end of the TSG study. There was not enough of the original batch of the California Standard cotton velvet to allow reevaluation of all 32 Series 100 cigarettes, so a subset of eight was chosen representing (a) ignition propensities that evenly spanned the entire range of ignition rates and (b) a distribution of values of each of the five design factors listed above. Those chosen were numbers 101, 103, 106, 108, 120, 129, 130, and 131.

Details of the results of the reevaluation can be found in [6], which has been included as Appendix A to this report. For three of the substrates, the data are consistent with the hypothesis of no change in the ignition properties of the cigarettes. However, there were some increases in ignition for cigarettes 101, 103, 129, and 130, with most being on the denim substrate. In the original evaluation, these four cigarettes tended to self-extinguish on the denim mockup, whereas in the reevaluation, mock-up ignitions tended to occur. The initial suggestion was that the change was due to deterioration of the denim fabric. However, Lorillard performed measurements of smolder proclivity using their published method [4], as well as weight, density and air permeability on the denim fabric, and determined that those properties had not changed with storage.

This result prompted a closer investigation of the two sets of ignition experiments. Three main differences were noted in the test methods:

- The original lab was not available for use in the reevaluation, so another test lab was used. The canopy hood in this lab had a slightly lower draw. This was not thought to be a serious problem because the smoke was being carried from the test chambers in a manner similar to the original study.

- A technician with no previous experience in ignition testing conducted the reevaluation tests. As a check, re-tests of cigarettes 101, 103, and 130 on the denim substrate were performed by the same operator who had performed the original TSG evaluation. The same tendency for more ignitions was noted.

- The original evaluation of the denim mockup was done in August, when the relative humidity in the test lab was 50 to 60 percent. The reevaluation was done in January and February, when the relative humidity was 30 to 40 percent. Other data indicate that a decrease this large can increase the number of ignitions. This suggests that the differences seen with certain cigarettes (101, 130) might be caused by this parameter. This particular substrate would be expected to be more sensitive to ambient humidity than the others in the TSG study since it virtually surrounds the cigarette with cellulosic materials--two cushions which form a crevice plus a cover fabric.

Table 1. Description of Series 100 Experimental Cigarettes

Experimental Designation	Packing Description				
	Tobacco Blend	Packing density	Paper Porosity	Paper Additive	Circum. (mm)
101 BNLC-21	Burley	Non-Expanded	Low	Citrate	21
102 BNLN-21	Burley	Non-Expanded	Low	No Citrate	21
103 BNHC-21	Burley	Non-Expanded	High	Citrate	21
104 BNHN-21	Burley	Non-Expanded	High	No Citrate	21
105 BELC-21	Burley	Expanded	Low	Citrate	21
106 BELN-21	Burley	Expanded	Low	No Citrate	21
107 BEHC-21	Burley	Expanded	High	Citrate	21
108 BEHN-21	Burley	Expanded	High	No Citrate	21
109 FNLC-21	Flue-Cured	Non-Expanded	Low	Citrate	21
110 FNLN-21	Flue-Cured	Non-Expanded	Low	No Citrate	21
111 FNHC-21	Flue-Cured	Non-Expanded	High	Citrate	21
112 FNHN-21	Flue-Cured	Non-Expanded	High	No Citrate	21
113 FELC-21	Flue-Cured	Expanded	Low	Citrate	21
114 FELN-21	Flue-Cured	Expanded	Low	No Citrate	21
115 FEHC-21	Flue-Cured	Expanded	High	Citrate	21
116 FEHN-21	Flue-Cured	Expanded	High	No Citrate	21
117 BNLC-25	Burley	Non-Expanded	Low	Citrate	25
118 BNLN-25	Burley	Non-Expanded	Low	No Citrate	25
119 BNHC-25	Burley	Non-Expanded	High	Citrate	25
120 BNHN-25	Burley	Non-Expanded	High	No Citrate	25
121 BELC-25	Burley	Expanded	Low	Citrate	25
122 BELN-25	Burley	Expanded	Low	No Citrate	25
123 BEHC-25	Burley	Expanded	High	Citrate	25
124 BEHN-25	Burley	Expanded	High	No Citrate	25
125 FNLC-25	Flue-Cured	Non-Expanded	Low	Citrate	25
126 FNLN-25	Flue-Cured	Non-Expanded	Low	No Citrate	25
127 FNHC-25	Flue-Cured	Non-Expanded	Low	Citrate	25
128 FNHN-25	Flue-Cured	Non-Expanded	High	No Citrate	25
129 FELC-25	Flue-Cured	Expanded	Low	Citrate	25
130 FELN-25	Flue-Cured	Expanded	Low	No Citrate	25
131 FEHC-25	Flue-Cured	Expanded	High	Citrate	25
132 FEHN-25	Flue-Cured	Expanded	High	No Citrate	25

Table 2. Ignition Propensity of Series 100 Experimental Cigarettes [3]

Cigarette Designation	Number of Ignitions in 20 Tests	Fraction of Ignitions
101	13	0.65
102	12	0.60
103	17	0.85
104	19	0.95
105	6	0.30
106	1	0.05
107	11	0.55
108	7	0.35
109	15	0.75
110	16	0.80
111	19	0.95
112	20	1.00
113	6	0.30
114	4	0.20
115	14	0.70
116	12	0.60
117	18	0.90
118	18	0.90
119	20	1.00
120	20	1.00
121	14	0.70
122	7	0.35
123	15	0.75
124	15	0.75
125	18	0.90
126	17	0.85
127	20	1.00
128	20	1.00
129	10	0.50
130	4	0.20
131	15	0.75
132	12	0.60

2. Series 500 Experimental Cigarettes

The remaining supply of several of the TSG cigarettes was insufficient for use throughout the present study, especially in the round robins. This led NIST to request from the cigarette industry a new lot of experimental cigarettes. Since the Series 100 cigarettes had shown a near-continuum of ignition propensities, the new Series 500 cigarettes were to be comparable in the five properties described earlier. Approximately 10,000 of each were supplied by the industry and placed in freezers until conditioned for test usage.

Since the need for the current project was specimens with a breadth of ignition propensities, **it was not necessary to assume, nor was it assumed, that the counterpart cigarettes would be identical.** Only a modest effort was made to characterize the new samples. A random selection of eight cigarette types to be used in the test method development was conditioned at 55 ± 5 % RH. Forty of each were weighed and the mean and standard deviation were determined. These weights and standard deviations for both series are shown in Table 3. Table 4 shows the weight and standard deviations provided by the cigarette industry for the Series 100 and 500 cigarettes. It should be noted that there are some significant differences in (a) cigarette weights between the two series in each table, (b) the weights in the two tables, and (c) the standard deviations in the two tables. The sources of these differences are not known.

Table 3. NIST Comparison of Series 100 and 500 Cigarette Weights

Cigarette Identity	Weight (mg)	Std. Dev. (mg)
101	831	14
501	826	19
103	835	36
503	824	17
106	640	9
506	592	17
108	565	40
508	588	15
120	1090	42
520	1065	27
129	836	47
529	845	32
130	841	7
530	842	30
131	959	22
531	844	22

The same eight cigarette types were also tested to ascertain that they would demonstrate a range of ignition performance and to gauge how useful the TSG data would be in estimating their performance. The cotton duck/polyurethane foam mock-ups were the same as those described below for use in the Mock-Up Ignition Test Method, and 24 replicates were performed on each. The new and old ignition data are shown in Table 5. Clearly, the cigarette/substrate combinations do show a range of ignition propensities suitable for intra- and interlaboratory evaluation of the methods being developed. There is a general similarity of the two data sets, although they do not correlate exactly. It was not determined whether the differences were due to variations in the cigarettes, materials, apparatus, or laboratory conditions. It should be noted that variations between the two limited data sets are essentially within the reproducibility of the Mock-Up Ignition Test Method assessed in this report (see below).

Table 4. Cigarette Industry Comparison of Series 100 and 500 Cigarette Weights

Cigarette Identity	Weight (mg)	Std. Dev. (mg)
101	873	5
501	840	3
103	882	10
503	841	0
106	613	5
506	615	3
108	612	5
508	612	6
120	1131	6
520	1104	1
129	846	5
529	853	3
130	862	4
530	849	2
131	936	1
531	855	4

Table 5. Comparison of Ignition Propensities for Series 100 and 500 Cigarettes

Series 100 Cigarettes			Series 500 Cigarettes		
TSG Cig. No.	Number of Ignitions	% Ignitions	TAG Cig. No.	Number of Ignitions	% Ignitions
106	1/20	5	506	9/72	13
130	4/20	20	530	0/72	0
108	7/20	35	508	24/72	33
129	10/20	50	529	12/72	17
101	13/20	65	501	70/72	97
131	15/20	75	531	47/72	65
103	17/20	85	503	71/72	99
120	20/20	100	520	72/72	100

B. Mock-Up Ignition Test Method

This section begins with a brief review of the past use of upholstered furniture mock-ups. It continues with a detailed discussion of the individual factors considered in the final design of this test method, which uses mock-ups to measure ignition propensity of cigarettes. The method itself is delineated in Appendix B.

1. Previous Use of Mock-Ups

As noted above, an upholstery mock-up is a reproduction of the upholstered furniture ignition problem. This has led to the widespread use of mock-ups in conjunction with the assessment of the vulnerability of upholstery materials to cigarette ignition. Much of this work is reviewed in reference [7]. Essentially all of the early work in this area was focused on the assessment of the cigarette ignitability of upholstery materials with a particular emphasis on fabrics. One standard test method for upholstered furniture ignition, NFPA 260, for example, uses a single cigarette type and a single type of polyurethane foam to test fabrics and divide them into classes dependent on the extent of smolder spread away from the cigarette coal [8].

More recently, the cigarette type has been varied to discern the extent to which its parameters affect mock-up ignition. Ihrig et al. [4], tested four cigarettes on mock-ups constructed from 33 commercial cellulosic upholstery fabrics of varied weight and construction; the underlying cushioning material was either cotton batting or a single polyurethane foam. The mock-up configurations included flat, 90° crevice and 20° crevice (a crevice configuration involves two separate foam-covered cushions brought together at the angle indicated). The principal cigarette variables were circumference and tobacco packing density. From a statistical analysis of their results, the authors concluded that the fabric variables (alkali metal ion content, weight and density) dominated the behavior of the ignition process; only the total radiative heat output of the cigarette had a significant impact of the likelihood of ignition. They also found that fabrics gave a graded ignition response (*i.e.*, other than 0% or 100% ignitions) only over a rather narrow range of properties.

In a subsequent study, Ihrig *et al.* [9] studied separately the impact of varying the characteristics of the polyurethane foam. Here only two cigarettes and three fabrics were used, and all results were for the 90° or 20° crevice mock-up configurations. The principal foam variable influencing mock-up ignitability was found to be air permeability. It is probable that the sensitivity to this parameter is greater in the crevice configurations used than it is in a flat mock-up. Once again, the sensitivity of the ignition behavior of the system was inferred to be greater for a mock-up variable (foam air permeability) than for the cigarette variable examined (radiative heat output per cigarette).

The potential impact of cigarette modifications on the ignition of upholstered furniture mock-ups may be underestimated in these studies in that the cigarette designs were not varied as much as those in the TSG study [3]. However, these studies do illustrate the point that the ignition or non-ignition of a mock-up is dependent on both the cigarette design and the mock-up materials. Rhyne and Spears [10] applied this point to actual furniture using the model developed in Ref. 9 and various assumptions about the distributions of fabric and foam materials in the real world.

As will be seen below, variation in the properties of the fabric used in the mock-up provides a useful means of discrimination among cigarette ignition propensities.

2. Fabric Considerations for a Mock-Up Test Method

The previous work revealed some of the advantages, sensitivities and limitations of mock-up testing for research purposes. However, the present program is the first extensive effort to pursue a standard test method for cigarette ignition propensity. Thus, comparatively little attention has been given in previous work to the issue of the long-term reproducibility of the ignition behavior such mock-ups produce.

The principal focus in this study of mock-up systems capable of long-term reproducibility has been the consistency of the fabric. It is the fabric which most closely interacts with the cigarette and whose ignition (when the substrate is a polyurethane foam) sets the stage for all subsequent behavior of the mock-up. Both chemical and physical features of a fabric influence its smolder propensity.

It has long been known that the principal chemical feature affecting the smoldering ignition propensity of a cellulosic fabric is its content of alkali metal and alkaline earth cations [11]. Sodium and potassium ions are particularly prevalent in such fabrics [4]. Potassium ions, in particular, are present naturally in cotton; sodium ions appear to be commonly used in fabric dying processes. Both are also introduced from perspiration and soiling [3]. These metal ions are present in the fabric in the form of organic and/or inorganic salts. It has not been generally appreciated in the past that the *anion* associated with the metal cation has a substantial influence on the effectiveness of the metal in catalyzing fabric smoldering. Thus, in reference [4] the total sodium and potassium ion content in 33 fabrics was reported along with fabric ignition temperatures and yarn "smolder proclivity" (total time an individual yarn from a fabric smoldered); the correlation between these two measures of smolder propensity and the total metal ion content showed a lot of scatter, possibly because the metal ions were present in a variety of salts.

The smoldering ignition propensity of a fabric is also influenced by its physical characteristics; this is particularly true when the ignition source is a cigarette. The influence of contact with the mock-up surface on the cigarette coal was examined to a limited extent in this study. It was apparent that the heat loss into the fabric can temporarily slow or even completely stop the smoldering process in the cigarette coal; the magnitude of the disturbance depends on the cigarette design and on the thermal capacitance of the fabric. The fabric thickness, density, heat capacity and thermal conductivity all play a role in determining this effective thermal capacitance. Thus, fabric structure needs to be closely controlled in any standardized material to be used in mock-up testing.

<u>Criteria Used to Identify Suitable Fabrics.</u> Discussions with representatives of the fabric and furniture industries made it clear that there is no practical way to characterize quantitatively the relative popularity of the thousands of upholstery fabrics used in the soft furnishings at risk to fire. If sales records are kept by individual fabric manufacturers or their customers, they are not publicly available. Therefore, identifying a set of test fabrics representative of the real-world was not a feasible undertaking and alternative approaches were pursued.

The ideas in the preceding paragraphs were blended with other considerations to arrive at the following selection requirements for suitable test fabrics:

- susceptibility to ignition from smoldering cigarettes, making the likely candidate fabrics to be cotton, linen, modacrylic and acrylic;

- differentiation of the ignition propensities of various types of cigarettes;

- capability to provide reproducible test results;

- ready availability now and in the future, with essentially constant cigarette ignitability in successive batches.

- manufacture such that their chemical and physical properties can be reproduced (inter- and intra-bolt);

- consistency of surface characteristics, so that surface contact between the cigarette and fabric surface remains constant along the length of the cigarette tobacco column and across the length and width of the fabric bolt;

- no preference for smoldering ignition in one orientation (*i.e.*, warp or weft yarns), making fabrics with similar warp and weft yarn construction preferable;

- freedom from finishes (*e.g.*, for flame retardancy, durable-press, or crush resistance), since (a) perfectly even finish surface characteristics and adhesion are difficult to obtain in commercially produced fabrics and (b) some finishes may promote or prevent smoldering ignition of the fabric; and

- weight in range representative of fabrics that are commonly used in the commercial upholstery fabric marketplace (0.17-0.85 kg/m^2; 5-25 oz/yd^2). Fabrics below about 0.34 kg/m^2 (10 oz/yd^2) tend to wear rapidly; those above 0.85 kg/m^2 (25 oz/yd^2) are very difficult to shape to an article of furniture.)

Air permeability of the fabric was not one of the chosen criteria for three reasons: (1) this parameter was found to be relatively minor in the statistical model of Ihrig et al. [4]; (2) there is reason to believe that the oxygen coming through the fabric is a minor contributor to the oxygen needs of the cigarette coal; see Appendix C; (3) the primary means of oxygen permeation through the fabric is believed to be diffusive, whereas air permeability measurements are based on air flow resistance.

The levels of cations in the fabric were also not included in the criteria. The original intention was to control this level by doping to a cation level which assured sustained smolder propagation; the cotton ducks that were ultimately used have such a cation level in their as-received state (see below).

To survey for appropriate fabric criteria and potential fabrics for use in a cigarette test method, NIST consulted with:

- research and test labs (California Bureau of Home Furnishings and Thermal Insulation, Department of Defense - Natick Textile Research Labs, Consumer Product Safety Commission),

- textile and furniture trade associations (American Textile Manufacturers Institute, American Furniture Manufacturers Association),

- textile mills (Glen Raven Mills, Mt. Vernon Mills, Graniteville Mills, J.B. Martin, and West Point Pepperell, Inc.),

- a textile distributor (Douglas, Inc.),

- NIST test method development staff, and

- a company which supplies standardized fabrics (Test Fabrics, Inc.).

Each of these parties has experience with either developing flammability test methods/standards or standardized fabrics or producing, using or distributing commercial fabrics. Each party was asked to list criteria important to developing a standardized fabric for test method use, describe problems associated with the production of standardized fabrics, and suggest possible fabric types for use in the test method anticipated here.

Cross-referencing the suggested practices and fabric types against the needed fabric characteristics noted above led NIST to the selection of cotton ducks as the candidate fabrics. These have a simple physical structure (plain weave) subject to control of weave details and air permeability, a long history of manufacture, conformance to a military specification [12], and at least limited usage as upholstery fabrics. They present a smooth surface to the cigarette coal, minimizing variations in heat transfer from the coal to the fabric. They are also made from a single component, raw cotton. Having no pile such as that in the fabric used for testing by the State of California ("California velvet"), they require no added finish to achieve a uniform physical appearance. These fabrics were thus judged to be excellent candidates for use in a mock-up method.

The physical properties of the 100% cotton fabrics examined in this study are summarized in Table 6; only a subset of these was ultimately utilized in the test method (Duck #4, #6 and #10). All were manufactured by West Point Pepperell Mills of West Point, Georgia (now known as Wellington Sears

Company)[1,2]. Since all are made from raw cotton (Texas, short staple) it is expected that their chemical composition is nominally similar. (The metal cation content was checked separately, as noted below.) The cotton was card cleaned using mechanical agitation only. No lubricants, surfactants or sizing were added to the cotton during the cleaning, carding, roving, spinning or the weaving processes. The yarns were made using open-end spinning frame technology. The fabrics are known as "greige" goods because they have no finishes or dyes.

Table 6. Specified Nominal Properties of Fabrics

FABRIC DESIGNATION	AREAL DENSITY	YARN COUNT (PER INCH)	YARN PLIES	AIR PERMEABILITY*
Duck No. 4 Style S/01400240	0.83 kg/m^2 (24.5 oz/yd^2)	31 x 24	4 x 4	5.1 - 10.2 x 10^{-3} m^3/s/m^2 (1 - 2 ft^3/min/ft^2)
Duck No. 6 Style S/01600230	0.72 kg/m^2 (21.2 oz/yd^2)	36 x 26	3 x 3	5.1 - 10.2 x 10^{-3} m^3/s/m^2 (1 - 2 ft^3/min/ft^2)
Duck No. 8	0.61 kg/m^2 (18 oz/yd^2)	34 x 27	3 x 3	5.1 - 10.2 x 10^{-3} m^3/s/m^2 (1 - 2 ft^3/min/ft^2)
Duck No. 10 Style S/01102020	0.50 kg/m^2 (14.7 oz/yd^2)	40 x 28	2 x 2	10.2 - 20.4 x 10^{-3} m^3/s/m^2 (2 - 4 ft^3/min/ft^2)
Duck No. 12	0.39 kg/m^2 (11.5 oz/yd^2)	46 x 35	2 x 2	20.4 - 30.6 x 10^{-3} m^3/s/m^2 (4 - 6 ft^3/min/ft^2)
Twill	0.52 kg/m^2 (15.3 oz/yd^2)	40 x 28	2 x 2	10.2 - 20.4 x 10^{-3} m^3/s/m^2 (2 - 4 ft^3/min/ft^2)

* Measured by Federal Method 5450 (contained in Federal Test Method Standard 191A, July 1978)

The chief differences in these fabrics should reside in their physical properties, since chemically they are raw cotton with comparable metal ion contents (see below). It is likely that the most important difference is the areal density, which varies by a factor of two. The potential heat sink effect to a cigarette coal thus varies by this same factor among these fabrics. The air permeabilities vary by a factor of three but, as will be seen, the mock-up configuration which was used is flat, and its ignitability should be relatively less sensitive to this parameter since more of the cigarette coal's periphery is exposed to ambient air. (Fabric permeability ranked fourth in order of importance as a controlling variable in the ignition of a flat mock-up in reference [4]. Fabric weight and total sodium/potassium ion content were the two dominant parameters.)

---

[1] The fabrics can be purchased from Wellington Sears Company, 3202 34th Street, Valley, AL 36854; telephone no. (205) 768-1222.

[2] Certain products or manufacturers are identified in this report in order to provide sufficient definition of procedures, equipment, and materials. In no case does such identification imply endorsement by the National Institute of Standards and Technology nor is the item identified necessarily the most appropriate for the purpose.

In anticipation of the fabric ignitability behavior discussed below, it is worth pointing out here that the ease of ignition of the cotton ducks in Table 6 is the opposite of what one might expect from previous literature results. The review of previous work [7] notes that cigarette ignition resistance decreases with increasing fabric weight. As will be seen below, the observable behavior of the fabrics in Table 6 is opposite to this trend; the heavy ducks ignite less readily than the lighter ducks. A plausible explanation of this is as follows.

The observed behavior in both situations (previous literature and here) is not the ignition event itself, which occurs close in to the cigarette coal, but rather the *sustained smolder spread* away from the cigarette coal (if and only if this spread can occur). The previous literature, with the possible exception of one experimental cigarette used in reference 4, is all based on commercial cigarettes which qualify as strong local ignition sources. The coal combustion for these cigarettes is sufficiently robust to overcome the heat losses to essentially the whole spectrum of fabric weights used in upholstered furniture; that is, they provide a sufficient heat flux to the fabric to ignite it locally in essentially all cases. However, among the commercial fabrics on which the previous literature is based, the heavier fabrics have a lesser surface-to-volume ratio, which yields a lesser heat loss rate and a greater tendency to propagate smoldering once it is locally initiated. Thus, given a strong igniter such as a commercial cigarette, a population of varying fabrics (having diverse levels of areal density, metal cation content and weave structure) will show a tendency for the observable part of the cigarette ignition process to be enhanced by increased fabric weight. The areal density or fabric weight effect will be most pronounced for those fabrics whose other parameters (metal cation content or weave structure) tend to be marginal in sustaining smolder propagation.

Here, however, the focus is shifted more specifically to *whether* local smoldering ignition of the fabric occurs. The cigarettes used are not necessarily strong igniters, but the cotton duck fabrics will smolder readily if ignited. Many of the experimental cigarettes used here are so disturbed by the heat loss they experience when in contact with the fabric that they go out. Others survive, but the coal is weakened in the area of contact with the fabric. Thus, in this case the transient heat sink effects of the fabrics are paramount. Heavier fabrics are greater heat sinks and therefore more ignition resistant.

<u>Additives as a Possible Means of Smoldering Ignitability Control.</u> Because commercial fabrics can show significant lot-to-lot variability in chemical and physical parameters, a substantial effort was made in the present study to develop a set of controlled fabrics. The cotton ducks in Table 6 were the basis for this development. As noted above, the cotton ducks have the necessary physical property control. As a means to render them completely specifiable with regard to cigarette ignition propensity, controlled doping with alkali metal and alkaline earth salts was investigated.

Appropriate salts must provide unambiguous self-sustained smolder propagation in the fabric when present above some minimum level. Above this minimum, they must also yield a differential ignitability response in the fabric when exposed to experimental cigarettes having differing ignition propensities (as judged by their behavior in the TSG study, reference 3). In practice, this last requirement probably translates into an ignition temperature which is in just the right range for some (not all) cigarettes to be able to induce in a fabric and which decreases continually with increased salt concentration. At the beginning of this study the identity of a suitable metal salt was unknown; and, as noted above, the important role of the anion was not known either. A variety of salts suggested by the limited literature in this field was examined:

- potassium chloride,
- potassium acetate,
- calcium acetate,
- sodium bicarbonate,
- mixtures of sodium borate with boric acid,
- potassium acetate with boric acid, and
- potassium acetate with diammonium phosphate.

All of these potential additives eventually were rejected because none could produce cigarette differentiation when present in the cotton ducks at levels sufficient to assure evenly propagating, self-sustained smolder. Furthermore, a problem with locally nonuniform deposition of the salts in the cotton ducks compounded the difficulty of the search and was not completely solved. Laundering and acid-washing of the fabrics prior to salt treatment proved insufficient to assure uniform penetration by the aqueous salt solutions. Commercial scrubbing followed by doping with commercial padding equipment probably could have resolved these difficulties, which may have been caused by natural waxes in the cotton.

Interestingly, the salts naturally present in raw cotton show no evidence in their smolder behavior of local non-uniformity problems, and tests showed that the unaltered fabrics in Table 6 could provide cigarette differentiation. Consultation with personnel at the USDA Southern Regional Laboratory [13] together with information from a standard reference text [14] indicated that the dominant salt in raw cotton is potassium malate. This salt is not commercially available. Limited studies with small quantities produced in our laboratory indicated that it could yield cigarette differentiation behavior similar to that seen with the cotton ducks in their "as-received" states. The non-availability of this salt, coupled with the lack of commercially scrubbed fabrics as hosts (even in small-scale laboratory studies) led to the termination of this approach to test fabric production.

Since the cotton ducks possessed all the desired properties of a controlled fabric for a mock-up based test method, including the desired cigarette differentiation in their as-received state, further development was pursued with these as-received cotton ducks as the fabrics of choice. Given this, it was necessary to assure that they could continue to meet the necessary criteria as to availability and invariant ignitability.

<u>Continued Availability of Cotton Duck Fabrics.</u> The simple plain or basket weave construction and desirable properties (high abrasion resistance, strong tear and tensile strengths) of cotton ducks make them highly sought-after products. For example, the U.S. Department of Defense has developed a number of specifications for cotton ducks which results in highly standardized fabrics. The military uses large quantities of these fabrics in products such as upholstery (camp seating slings), backpacks, tenting, sandbags, and medical stretchers. Commercially, cotton ducks are commonly used as an upholstery fabric in director's chair canvas slings. They have also been used in upholstered furniture, but this use is driven by home fashion trends. Currently they are featured as upholstery fabrics in a number of mail order and furniture periodicals [15].

As a result, these fabrics are produced in bountiful supply by textile companies throughout the world. In fact, cotton duck fabrics have been produced continuously for more than 200 years. There are approximately 34 million m^2 of cotton ducks (greater than 50% cotton content) sold annually in the United States. This information provides a high degree of assurance that cotton ducks will be readily available and produced in a consistent and standardized manner.

Metal Ion Content Over Time. Since cotton ducks are made from raw cotton, their content of alkali metal and alkaline earth ions is potentially variable with soil, fertilization and growth conditions. Blending of raw cotton from various regions (of Texas) and crop years tends to counteract this variability. Recognizing the potential problems here, NIST sought to develop information on the extent of variability of cation content in cotton ducks. This process was greatly simplified by determinations that:

- the alkali metal ions are comparable in smolder promotion tendency and much more potent than the alkaline earth cations [16] and

- potassium ions are present in dominant concentrations in the cotton ducks and the relative fractions of the other metal ions varied little (Table 7).

The premise adopted was that the potassium ion concentration is the determining chemical factor in ignition susceptibility of these fabrics. The malate anion is equally important in setting the general level of activity of the potassium. Since this is the dominant anion in cotton [13] it is expected to correlate with the potassium level, barring any major genetic modifications to future cotton strains.

NIST then worked with West Point Pepperell (WPP) to examine the long-term reproducibility of the potassium ion content of the ducks. WPP staff utilized the NIST sample extraction technique (Appendix D) and atomic absorption spectroscopy to analyze samples from their mill for potassium ion content over a period of 4 months. (A reorganization of the company prevented a longer analysis period.) The results are shown in Table 8. Each duck was sampled in three locations during one day of each month reported; the standard deviations shown are for these three measurements. There is only one case (Duck #6 in June, 1992) of highly variable results. Otherwise the spatial variability on a given day is ± 6% or less. The long-term variation tends to be greater but, except for the one case of Duck #6 (June, 1992), the variation is not very large. Duck #8 shows the greatest variation, a 23% increase from 4700 to 5800 ppm, from April to May, 1992.

Table 7. Cation Content of Fabrics Used in the Preliminary
and Main Interlaboratory Studies

Duck Number-Bolt Number	[Cation] (ppm ± one Standard Deviation)			
	Na^+	K^+	Mg^{+2}	Ca^{+2}
4-46*	<20	4575±133	607±19	691±26
4-48*	<10	4243±37	582±6	683±5
4-50	<15	4477±75	567±12	607±56
4-52	<20	4546±125	566±29	575±44
4-54	<25	4528±55	558±5	569±21
4-56	<20	4510±44	564±3	564±16
6-65*	<20	5667±185	653±13	748±13
6-67*	<35	5900±107	656±12	727±25
6-69	<45	4573±257	573±19	575±37
6-71	<30	5742±102	633±19	690±37
6-73	<15	4439±143	578±14	650±11
10-57*	<50	4445±88	607±9	708±16
10-58	<60	4214±71	580±10	691±17
10-59*	<20	4422±94	605±14	698±22
10-61	<60	4224±111	590±12	665±3
10-63	<70	4069±162	575±19	663±33

* Used in preliminary interlaboratory study; otherwise used in main interlaboratory study

Table 8. Potassium Content of West Point Pepperell Cotton Ducks Over a Four Month Period

TIME	DUCK NO.	POTASSIUM LEVEL (ppm)
April, 1992	4	5200 ± 220
" "	6	5200 ± 260
" "	8	4700 ± 200
" "	10	5500 ± 190
May, 1992	4	5400 ± 170
" "	6	5600 ± 80
" "	8	5800 ± 230
" "	10	6000 ± 200
June, 1992	4	5800 ± 270
" "	6	8200 ± 2200
" "	8	5600 ± 50
" "	10	5700 ± 35
July, 1992	4	6000 ± 170
" "	6	5500 ± 340
" "	8	5800 ± 250
" "	10	6000 ± 24

Effect of Ion Content Variation on Mock-Up Ignitability. Table 9 shows the results of limited testing (5 replicates, 3 cigarette types) using ducks #4, #6 and #10 from the analyzed lots described in Table 8. The ignition propensities are comparable despite the noted variations in the potassium ion content of the fabric. The widest variations in potassium content were not included in this testing.

Table 9. Sensitivity of Ignition Susceptibility to K^+ Content in Fabrics; 5 Replicates

Fabric	[K^+] (ppm)	Percent Ignition for Cigarette		
		#506	#529	#503
Duck #4	5400	0	0	100
"	6000	0	0	100
Duck #6	6000	0	20	100
"	8200	0	0	100

Limited testing was also done on a #8 duck from another manufacturer, obtained through the American Textile Manufacturers Institute (ATMI). Analysis showed this fabric to contain ≈ 100 ppm of sodium, 3500-5100 ppm of potassium, 450 ppm of calcium, and 320 ppm of magnesium. This was compared to WPP duck #8, which Table 8 shows to contain 4700 to 5800 ppm of potassium. The other, less critical metals were not greatly different from those in the WPP duck (Table 7). Six TSG cigarettes of differing ignition propensity again showed comparable ignition propensities on the two ducks (Table 10). (Comparable, as used in this context, means that any differences in ignition propensity were below the typical levels of scatter seen in these tests; this issue is discussed more thoroughly in the context of the round robin studies below.)

Table 10. Ignition Susceptibility of Different #8 Cotton Duck Fabric Samples
(Percent Ignition in Six Replicates)

Cigarette Number	WPP Duck	ATMI Duck
106	0	0
114	0	0
108	0	0
129	17	33
101	100	100
120	100	100

The cation content of all fabrics used in the interlaboratory testing described below was monitored along the length of the fabric bolts used by the method described in Appendix D. Depending on the bolt length, anywhere from 3 to 10 samples were taken along the length of a given bolt and analyzed for sodium, potassium, magnesium and calcium content. A summary of this cation content is shown in Table 7. The numbers are the average of the samples taken on each bolt of fabric (± one standard deviation). Appendix D contains the cation content for all the individual samples tested. The most variable fabric is duck #6, with potassium levels ranging from about 4400 ppm to 5700 ppm in the bolts used in the main interlaboratory study (described below in Sect. B.8). This is a substantial range (*ca.* 30% referred to the smaller number), but it did not result in any extraordinary variability in the interlaboratory results obtained with this duck. The implication thus is that variations in metal cation content comparable to those seen in Table 7 (which in turn are comparable to those seen over the four-month period shown in Table 8) are not detrimental to the reproducibility of the mock-up test method discussed below.

The potassium levels in Tables 7 and 8 may seem high compared to many (not all) of the 33 commercial fabrics analyzed in the work of Ihrig [4]. However, this misses the role of the anion in shifting the catalytic effectiveness of the cation. Unfortunately, anion measurements were not made in reference 4. Thus, the relation of those results to the present levels, in terms of ignitability enhancement, is unknown.

For the best long-term reproducibility it is preferable that the potassium ion levels not be in a domain where the ignition behavior is sensitive to small changes in potassium level. The above results indicate that the potassium levels in the cotton ducks are indeed well above the sensitive region. The sensitive region for potassium acetate, noted in cigarette industry studies, was *ca.* 2000 ppm.

Physical Variability of Cotton Duck Fabrics. Areal density is believed to be the most important physical property affecting ignition susceptibility of the cotton ducks. The variability of this property along the length of the fabric bolts used in the interlaboratory studies described below is indicated in Table 11. The standard deviations and coefficients of variation are based on five samples from along the length of each bolt.

Air permeability measurements performed in accord with ASTM Method D 737-75 [17] were made on samples from several of the same bolts by the United States Testing Company. Five samples from each bolt were measured; the results (± the standard deviation) are shown in Table 12. The test method, apparatus, and pressure drop were fundamentally the same as that used to set the nominal air permeability specifications in Table 6. This small degree of physical variability in the cotton ducks was further reinforcement of the appropriateness of these fabrics for use in the interlaboratory study.

Also shown in Table 12 are the measured air permeability values for the three principal fabrics used in the TSG study [3]. The large variability of the Splendor fabric is the result of one particular sample; a coefficient of variation closer to that of California Velvet typified the other three samples measured here. It is of interest to note that the TSG fabrics have permeabilities that are ten to twenty times higher than the cotton ducks used here. This will not preclude similar types of ignition behavior from being exhibited by the two groups of fabrics, as will be seen below.

Table 11. Measured Areal Densities of Fabrics Used In Interlaboratory Study

Duck Number-Bolt Number	Areal Density (g/m^2)	Coefficient of Variation (%)
4-48	820 ± 17	2.0
4-52	806 ± 25	3.1
4-56	803 ± 14	1.7
6-67	712 ± 9	1.3
6-71	705 ± 18	2.6
10-58	506 ± 18	3.5
10-63	496 ± 6	1.1

Table 12. Measured Air Permeability of Fabrics Used in Interlaboratory Study and in TSG Study

Duck Number-Bolt Number or Fabric Name	Air Permeability*	Coefficient of Variation (%)
4-52	(8.89 ± 0.15)x10^{-3} m^3/s/m^2 (1.75 ± 0.03 ft^3/min/ft^2)	1.7
4-56	(8.74 ± 0.91)x10^{-3} m^3/s/m^2 (1.72 ± 0.18 ft^3/min/ft^2)	10.5
6-67	(5.54 ± 0.25)x10^{-3} m^3/s/m^2 (1.09 ± 0.05 ft^3/min/ft^2)	4.6
6-71	(5.54 ± 0.15)x10^{-3} m^3/s/m^2 (1.09 ± 0.03 ft^3/min/ft^2)	2.8
10-58	(10.72 ± 0.71)x10^{-3} m^3/s/m^2 (2.11 ± 0.14 ft^3/min/ft^2)	6.6
10-63	(11.53 ± 0.61)x10^{-3} m^3/s/m^2 (2.27 ± 0.12 ft^3/min/ft^2)	5.3
Splendor	0.12 ± 0.04 m^3/s/m^2 (24.1 ± 7.7 ft^3/min/ft^2)	32.0
Blue Denim	(6.81 ± 0.30)x10^{-2} m^3/s/m^2 (13.4 ± 0.6 ft^3/min/ft^2)	5.0
California Velvet	0.12 ± 0.01 m^3/s/m^2 (23.2 ± 2.5 ft^3/min/ft^2)	11.0

* Data obtained by United States Testing Company using ASTM D 737-75

3. Other Mock-Up Materials

Two other expendable materials are used in the mock-up method. The principal one is a polyurethane foam which is used to mimic the typical cushioning material in upholstered furniture. A second material is a polyethylene film used between the fabric and foam in one mock-up configuration for reasons explained below.

<u>Polyurethane Foam.</u> The polyurethane flexible foam used in these test method development studies had the same formulation as that used in the TSG study. The foam is based on a polyether polyol and TDI; the manufacturer's (Vitafoam, Inc., High Point N.C.) designation is 2048.[3] It has an indent flexural rating of approximately 21.8 kg (48 lbs) and a nominal density of 32 kg/m^3 (2.0 lb/ft^3). The nominal air permeability (ASTM D3574 [18]) is 2.0 x 10^{-3} m^3/s (4.25 ft^3/min). The foam is representative of foam products used in the residential furniture market.

The sensitivity of the cigarette ignition process to foam properties was examined by substituting another common upholstered furniture foam. This foam had a similar TDI/polyether formulation, but a nominal density of 24 kg/m^3 (1.5 lb/ft^3) and a nominal air permeability of 2.4 x 10^{-3} m^3/s (5.0 ft^3/min). Flat mockups were made with duck #8 and the two foams. TSG cigarettes nos. 108 (7/20 TSG ignitions), 129 (10/20 TSG ignitions), 102 and 116 (both 12/20 TSG ignitions) were tested on the mockups using six replicates per cigarette/mock-up condition. See Table 13.

Table 13. Sensitivity of Ignition Susceptibility to Foam Properties
(Percent Ignitions in Six Replicates)

Cigarette Number	Ignitions (Heavier Foam)	Ignitions (Lighter Foam)
108	50	33
129	50	17
102	100	100
116	100	100

Since the foam density variation in this experiment is substantially larger than would occur within any well-specified foam batch (± 5%) and since the effect here was small, it was concluded that the role of foam property variations (within nominally similar formulations) is minimal. It should be sufficient to specify the general formulation and nominal density.

From consulting with experts on polyurethane foams, it was determined that the greatest (± 5%) variation in foam density occurs vertically in a bun. The air permeability varies similarly; see Table 14. In the interlaboratory testing described below, the foam samples were varied randomly from top

---

[3] The foam was obtained from TEDCO, 2335 W. Franklin Street, Baltimore MD 21223; telephone no. (410) 945-6158. TEDCO identifies this foam as style #2045.

to bottom of the bun. As will be seen, the impact on the inter- and intra-lab variability was at an acceptable level. This means that the density and permeability range typical of current foam manufacturing practice are an acceptably small source of scatter in mock-up ignition behavior.

Table 14. Measured Air Permeability of Polyurethane Foam By ASTM D 3574
(Average of 3 to 4 samples at each location.)

Foam Bun	Location	Air Permeability
A	Middle	$(1.83 \pm .02) \times 10^{-3}$ m^3/s $(3.89 \pm .04$ ft^3/min)
A	Top	$(2.00 \pm .03) \times 10^{-3}$ m^3/s $(4.24 \pm .06$ ft^3/min)
B	Middle	$(1.80 \pm .01) \times 10^{-3}$ m^3/s $(3.82 \pm .03$ ft^3/min)
B	Top	$(2.01 \pm .02) \times 10^{-3}$ m^3/s $(4.26 \pm .03$ ft^3/min)

Polyethylene Film. In one of the mock-up configurations ultimately included in the test method described below, a polyethylene film was placed between the fabric and foam as an additional heat sink to make the mock-up more ignition resistant. Inadvertently, different films were used in the preliminary and the main interlaboratory studies described below. Table 15 lists the properties of the two films.

Table 15. Properties of Polyethylene Films Used in Conjunction with Duck #4

Property	Poly-America, Inc. (Preliminary RR)	Warp Bros, Inc. (Main RR)
Thickness (mm)	$0.15 \pm .007$	$0.13 \pm .005$
Density (g/cm^3)	0.79	1.15
Areal Density (g/cm^2)	0.012	0.015
Melting Points (°C)*	118, 124	115, 122

* Two distinct peaks for crystalline regions were found for each polymer film.

As will be seen below in comparing the preliminary and main interlaboratory results, these property differences (most likely the areal density difference) were sufficient to yield differing ignition propensity measurements on two cigarettes in the interlaboratory studies. The film to be used in the test method is specified similar to the one manufactured by Warp Brothers, Inc. under the trade name Poly-Film; it was obtained from Read Plastics, Rockville, MD 20852. The reason for this preference emerges from the interlaboratory studies described below.

4. Mock-Up Configuration

Several issues were considered in deciding how the mock-up assemblies were to be configured. These affect the degree of replication of the real-world situation, ease of fabrication, and reproducibility of test results.

The first issue concerns fabric/foam contact. Wrapping the fabric around the foam (totally or partially), as done in earlier studies, makes it difficult for the test operator to obtain reproducible, even and constant tension of the fabric over the foam. The resulting variation in surface contact between the fabric and foam changes the local thermal capacitance of the mock-up, which in turn affects its susceptibility to ignition. This is especially important for the cotton ducks, which are extremely flat and maintain very good surface contact with the foam in a flat configuration, but for which side wrapping of the fabric around the foam would produce a significant surface contact problem.

A second issue concerns whether the mock-up should mimic a crevice or a flat area of upholstered furniture. The greatest realism would doubtless come in some degree of crevice configuration. However, the crevice design introduces reproducibility problems. Accurate placement of the two cushions to form the crevice is important so that the intersection line is even and repeatable. This difficulty is compounded by the sensitivity of a cigarette's ignition propensity to its placement relative to both surfaces. Tests at CSIRO in Australia have indicated that the outcome of a crevice test (ignition or nonignition) can be heavily influenced by how firmly the operator places the cigarette in the crevice [19]. This introduces a potentially strong operator dependence that is undesirable.

Third is the desired degree of ignition susceptibility of the particular mock-up to the heat produced by the cigarette. In the TSG full-scale furniture tests [3, 3], the commercial cigarette, a strong igniter, generally showed a higher fraction of ignitions in the crevice configuration. Apparently the cigarette coal generated enough heat to overcome the high thermal capacity of two fabric surfaces and the restricted oxygen flow to the combustion zone. The four experimental cigarettes, with lower bench-scale ignition propensities and presumably lower heat transferred, showed no consistent trend between crevice and flat configurations. Various crevice substrates in the full-scale chairs produced higher, similar or lower fractions of ignitions than the flat systems comprised of the same fabric and padding. These results suggest that the flat configuration might better differentiate among cigarettes of high ignition propensity than the crevice; on the other hand, Ihrig et al. [4], using four cigarettes on thirty fabrics, found the crevice to discriminate among their cigarettes while a flat mock-up did not. For the cotton ducks used in this study, limited experiments were performed to see if a crevice mock-up would aid in discriminating among the high ignition propensity cigarettes. The crevice mock-up was found to be more ignitable and thus not helpful in seeking the desired discrimination. For cigarettes of lower ignition propensity, there is no clear advantage of either configuration.

A fourth consideration is the surface size of the mock-up. This should be large enough to accommodate any reasonable length cigarette, while being small for ease of maintaining uniformity of contact between the fabric and the lower layer(s) of the substrate.

For these reasons, it was decided to test in only the flat configuration. In addition, a square, flat brass frame (20 cm outer edge, 2.54 cm wide) was developed for placement on top of the fabric to assure that it remained in excellent contact with the foam below. The use of the frame is distinctly more reproducible than anchoring the fabric edges with pins. The frame also guarantees that the cigarette is placed in the same mockup location from test to test. The hot cigarette coal is placed in the center of the mockup and the non-ignited tip (filter) of the cigarette is oriented toward one of the right-angled corners of the frame.

The mockup was enlarged, compared to the mockups used in the TSG study, to 20.3 cm x 20.3 cm (8" x 8"). This provides an ample-sized mockup for almost any cigarette length and eliminates the need to determine the warp or weft orientation of the fabric with respect to mockup orientation. Placing the cigarette on the mockup at a 45° angle assures that the smoldering cigarette tobacco column will make equal contact with the warp and weft yarns of the fabric.

With this flat configuration, consisting simply of a square of cotton duck held in good contact atop a square of polyurethane foam (5.1 cm thick), a series of screening tests was performed to determine the degree of ignition propensity differentiation provided by the various fabrics. Table 16 summarizes the results.

Table 16. Percent Ignitions on Various Substrates for Selected Cigarettes
Flat Configuration; 4 to 6 Replicates

Fabric → Cigarette # and TSG Ignitions ↓	Duck #6	Duck #8	Duck #10	Duck #12
106 (1/20)	0	0	33	67
114 (4/20)	0	0	33	67
113 (6/20)	0	0	50	100
108 (7/20)	17	0	50	100
129 (10/20)	25	50	67	100
101 (13/20)	100	100	100	100
120 (20/20)	100	100	100	100

Table 16 shows that these fabric/foam mock-ups do provide varying degrees of differentiation of the cigarettes. Ducks #6 and #8 were similar to each other. Duck #10 was more readily ignited. Duck #12 (and the twill fabric in Table 6) provided only minimal differentiation among the weakest igniting cigarettes. Duck #4, when assessed with a different set of TSG cigarettes (114, 108, 107, 101, 124, and 125), showed a transition from non-ignition to ignition not greatly different from that of Duck #6.

It was also desirable to have at least one mock-up which would be resistant to all but the most ignition prone cigarettes (*e.g.*, TSG rankings of 15/20 through 20/20). It is well known that polyester battings used in upholstered furniture act as a heat sink and absorb the energy from a smoldering cigarette. This suggested the use of a similar concept, the use of a thin, high density heat sink material in better thermal contact with the fabric than is the case with low density batting. This was incorporated into a mock-up consisting of the heaviest fabric, duck #4, and a thin thermoplastic film to serve the role of added heat sink. The Poly-America film listed in Table 15 served this role. Generally, cigarettes with a TSG test result of 16/20 ignitions and above are required to ignite this substrate though there was at least one anomaly (cigarette 102, with a TSG rating of 12/20 gave six ignitions in six replicates). The Warp Brothers PE film used in the main round robin proved even more ignition resistant.

5. Enclosure Design; Air Flow Considerations

The reason for enclosing the mock-up during a test is to isolate it from random, uncontrolled air currents which could lead to non-reproducible ignition behavior. A very simple open-top enclosure was utilized in the previous study [3]. This was reasonably effective, but it did not completely prevent eddies induced by the laboratory ventilation system from causing occasional visible disturbances of the smoke plume issuing from a cigarette on top of a mock-up. The flow disturbances were measured at up to 8 cm/s. The data from those mock-up tests correlated well with those from full-scale tests in which the air flow disturbances were similarly random (in time and orientation) but of somewhat greater magnitude (12-13 cm/s) [3].

The mock-up enclosure used in the present study is a modification of that designed by the cigarette industry for their own round robin testing. Figure B-1 in Appendix B shows a schematic of the enclosure and the associated smoke exhaust hood. The flow in the neighborhood of the cigarette is sufficiently low that the smoke plume rises totally undisturbed (visually) up into the chimney. Since the cigarette plume must act as a weak pump carrying some air out of the box, some replacement air must flow down the outer portions of the chimney, but its velocity is too low to measure. The oxygen level at the height of a burning cigarette drops no more than 0.1 to 0.2 % (below normal ambient levels) when a cigarette burns its full length in this box.

The cigarette industry has expressed concern about the role of ambient air flow and its potential ability to modify the ignition propensity of cigarettes. Changes of greatest concern would manifest themselves as alterations in the rank ordering of the cigarettes' ignition propensities at different air velocities. Of lesser concern is the potential for all ignition propensities to increase uniformly. Assessment of any changes in ignition propensities must consider the reproducibilities of both the study that generates such information and of the cigarette ignition test methods themselves. The former has not been addressed; the latter is discussed below in light of the interlaboratory study results. For the present, it is important to note that shifts in relative ignition propensity must be substantial (*i.e.*, 35% or more) to be judged significant. Cigarette industry staff have made several presentations of their studies of the air flow effects on ignition propensity. The most thorough and meaningful of these, in light of the above caveat, is discussed here.

In reference [20], Adiga *et al.* report on the effects of steady, low velocity flows impinging on cigarettes in the same direction as that in which the coal is moving. (This head-on flow impingement is the worst case with regard to impingement angle [21]. The steady, uni-directional nature of the flow can also be expected to yield a greater impact on the cigarette coal than does a randomly fluctuating flow that includes some flow reversals.) The peak flow velocity used there (5 cm/s on their "breeze tunnel" centerline) gave a flow velocity on the cigarette centerline of approximately 1 cm/sec (4 mm from the wall surface). This is about the same as the average buoyancy-induced velocity level reported by R. Flack (in a study for the cigarette industry) in the crevice region (4 mm from surface) of a chair previously heated by a *ca.* 37 °C heater simulating a person [22]. These real chair results also showed substantial flow fluctuations, including some flow reversals. While these are very low velocities, they are comparable in magnitude to those measured very near the top of a cigarette coal during natural smolder when mounted horizontally in free space [23]. Presumably the presence of a horizontal surface below the cigarette coal lowers the local plume velocity even more and renders it susceptible to alteration by small ambient velocities.

The impact of a flow disturbance on the cigarette coal is most likely to be one of increasing the coal temperature somewhat since oxygen transport to the coal will be enhanced. Heat losses will also be somewhat enhanced, but this effect should be smaller since the radiant component is not directly affected. The magnitude of any change in the coal temperature is not readily estimated, however, even from an ignition model because the mass transfer processes in the critical region of contact between coal and fabric are very complex. The impact on the fabric ignition process itself (*i.e.*, the runaway acceleration of fabric char oxidation reactions) may not be negligible. This runaway is somewhat retarded by oxygen depletion below the coal [3], and air flow could affect this.

The overall consequences of very low ambient velocities such as were noted above are ambiguous at present. The impact of disturbing the air in the NIST enclosure was examined experimentally. A small fan of the type used to vent electronics cabinets was mounted in one corner of the enclosure at mid-height. The fan speed was controlled with a variable transformer and its RPM was set with precise repeatability using a stroboscope. The fan blew upward so as to effect throughout the enclosure volume a large, recirculating eddy-like flow which passed over the cigarette atop a mock-up with the flow generally impinging head-on. The flow velocity fluctuated from 4 to 13 cm/s (uni-directional), blowing the smoke plume over at an angle that varied from 30° to 90° off vertical.[4] Even though some smoke accumulated in the enclosure in these circumstances, the oxygen level at the height of the cigarette did not drop more than 0.2% below ambient except when mock-up ignition was well along. Table 17 shows there was no significant effect of this flow.

When the fan RPM was doubled, yielding flow velocities that fluctuated in the range from 10 to 25 cm/s, cigarettes 108 and 508 did respond with a significant increase in the number of mock-up ignitions. Cigarettes 106, 130, 506, 508 and 529 did not; the other cigarettes all yielded essentially 100% ignitions under all conditions.

---

[4] These flow velocities must be regarded as approximate since they were at the low end of the capability of the anemometer used. The plume behavior was very clearly altered over its full height, however.

Table 17. Effect of Air Flow Disturbance on Cigarette Ignition Propensity Duck #6, Percent Ignitions

Cigarette # (TSG Ign. Frac.)	Replicates	No Flow	Flow	Double Flow
106 (1/20)	4	0	0	0
130 (4/20)	4	0	0	0
108 (7/20)	4	25	25	75
102 (12/20)	4	100	100	100
121 (14/20)	4	75	100	100
109 (15/20)	4	100	100	100
128 (20/20)	4	100	100	100
506	16	0	0	0
508	16	0	0	56
529	16	12	25	19
530	16	0	0	0

Adiga *et al.* [20] used the Series 500 cigarettes and cotton ducks stated to be comparable to those used here. Their polyurethane foam was 25% lower in density than that used by NIST; but, as noted above, we have found little effect of such a density difference. They also found a rather minimal response from cigarettes placed atop duck #6, although cigarettes 530, 505 and 529 did show some ignitions (10-30 %) at a steady, head-on airflow velocity of approximately 1 cm/s (cigarette centerline); with no flow these three cigarettes gave no ignitions. The lighter cotton ducks (#8, #10, #12) showed an increasing response to the same air flow, with the response being greatest for the lightest duck. In all cases, however, while the absolute number of ignitions went up, the relative ranking of the tested cigarettes remained similar to that seen with their TSG analogs. This type of result, an upward shift in number of ignitions with small changes in relative cigarette ignition propensity rankings, implies that testing with or without an ambient flow would produce little practical difference. In assessing results of this type one has to bear in mind the degree of reproducibility of the test and the limits this imposes on the ability to make distinctions in cigarette ranking. The reproducibility of the mock-up test method developed here is discussed in the context of the interlaboratory study below.

Adiga *et al.* [20] also examined the influence of air flow on the ignition behavior of the Series 500 cigarettes with two other fabrics, a blue denim and California Standard velvet. These are nominally the same as two of the fabrics used in the TSG study, except that they were doped with potassium acetate in this study to enhance their ignitability.[5] The doped California velvet proved to be too readily ignited by most of the Series 500 cigarettes to provide much information on air flow effects. The behavior on the blue denim was more complex. There was an increase in ignitions as the potassium level was increased, even in the no-flow case. At any given level of potassium, the presence of a steady air flow (*ca.* 1 cm/s at the cigarette centerline) enhanced the number of ignitions still further. The most distinctive anomaly in all of this is the observation that three of the cigarettes [505 (BELC-21), 506 (BELN-21), and 508 (BEHN-21)] showed a relatively stronger response to the air flow, and this tended to alter their ranking substantially relative to the other cigarettes tested. These are cigarettes whose TSG analogs exhibited low ignition propensities. Evidently, in the presence of the particular air flow conditions of this experiment, these cigarettes on this fabric lose their diminished ignition propensity and tend toward the behavior seen with high ignition propensity cigarettes. A physical explanation for this is lacking at this time. The extent to which this result would carry over to the real world is also not known at this time. As noted above, greater flow differences between mock-up and chair tests in the TSG study did not preclude a good correlation between the two types of tests.

In view of the information at hand, it has been judged appropriate to select the no-imposed-flow case as preferable since it clearly is simplest and, on balance, seems quite relevant to the real world. In the real world, the orientation of any flow relative to the cigarette coal is unknown but is probably random; it will depend on where and in what orientation the cigarette happens to fall. Many ignitions may occur down in a crevice-like crack, such as is formed by the seat cushion and the side of the chair; and the air flow there is likely to be very small (smaller than the values measured by R. Flack [22]). Thus, even cigarette designs such as those noted above as having lost their low ignition propensity in some particular sets of circumstances are expected to exhibit low ignition propensity in many real world conditions. Should more information on the response of cigarettes to real world conditions be developed in the future, it may be appropriate to supplement the no-imposed-flow test behavior with other data.

6. Test Variables

In order to optimize the test method specification, a list of parameters was compiled, prior to finalizing the method, with advice from the Technical Advisory Group, to identify possible sources of test variability (Table 18). These were classified by source: substrate type, test environment, test operator and test procedure. Based on the extant data at the time of initial list compilation, each parameter was assigned by NIST a sensitivity level that indicated its possible impact on the test outcome. For a standardized test method, it is desirable to have as many variables as possible determined to be "not sensitive."

---

[5] In the TSG study the California velvet was used over cotton batting in a flat mock-up configuration. The blue denim was used in a crevice mock-up configuration with a cover cloth over the cigarette. Neither ignites readily in a simple flat mock-up configuration over polyurethane foam as used by Adiga *et al.*

Based on the various experimental results described above and careful, detailed specification of the test procedure, NIST subsequently moved several of the variables in the "B" and "C" columns to the "A" column. These included:

- additives and impurities of the materials and their physical properties;

- fabric tension, retention method, and configuration; and

- mockup location in the box.

Others were assigned as variables to be assessed during the interlaboratory study:

- the operator variables,

- materials conditioning, and

- relative humidity and temperature in both the conditioning and test rooms.

The series of items under "cigarette ignition procedure" was resolved based upon data from NIST and the cigarette industry. These studies combined to establish a procedure that had minimal impact on ignition propensity:

- ignition by a gas lighter with a fixed flame size,

- a cigarette pre-burn, in the vertical orientation, to a length of 15 mm subsequent to ignition and prior to cigarette placement on the substrate,

- transport in a vertical orientation of the cigarette to the test chamber, so as not to dislodge the ash.

Since no significant changes in ignition propensity had been observed during the course of this study, it was presumed that "fresh" substrate materials did not age substantially over a year.

Table 18 is instructive in that it indicates the large number of variables which must be considered and controlled in order to assure a reproducible test outcome. Most are handled in a prescriptive manner by restrictions on materials and by a very explicit test procedure.

Table 18. Estimated Sensitivity of Mock-Up Test Outcome to Test Variables

A = Not sensitive if carefully controlled; B = Expected to be sensitive; C = uncertain of sensitivity

VARIABLE	A	B	C
SUBSTRATE			
Fabric			
fiber content	X		
additives		X	
impurities		X	
existence and variation in backcoating	X		
existence and variation in fiber coating	X		
yarn twist			X
warp & fill count	X		
air permeability			X
weave type	X		
pile depth	X		
areal density		X	
Foam			
air permeability			X
chemical formulation			X
age		X	
thickness	X		
additives		X	
inorganic content		X	
cell size			X
density		X	
Mockup			
dimensions	X		
fabric tension		X	
randomization of materials	X		
# sides covered by fabric	X		
fabric retention method (e.g., pins)		X	X
configuration (crevice, flat...)		X	
TEST ENVIRONMENT			
enclosure size	X		

VARIABLE	A	B	C
enclosure materials	X		
external air flow	X		
internal air flow	X		
mock-up location in box		X	
relative humidity		X	
temperature			X
OPERATOR			
experience level		X	
glove use in handling mock-ups	X		
mechanical handling of fabric		X	X
handling of cigarette	X		
cigarette placement on mock-up			X
ID of cigarettes	X		
TEST PROCEDURE			
allowed cigarette shelf life	X		
allowed materials shelf life			X
cigarette conditioning			X
mock-up conditioning			X
retrieval of components for test	X		
cigarette ignition procedure			
cigarette smolder line		X	
draw rate on cigarette		X	
ignition time and flame location		X	
movement to test box		X	
orientation of cig. during free burn		X	
ash retention		X	
placement of cig. on mock-up	X		
door closure speed		X	X
definition of ignition	X		
number of replicates	X		

7. General Description of Mock-Up Ignition Test Method

This test method depends on seven components which are considered to be critical:

- a test operator skilled in basic laboratory techniques,

- an environmental room/chamber for preconditioning the cigarettes and mock-up assemblies,

- an environmentally-controlled test room,

- a cigarette lighting apparatus,

- a test chamber,

- a furniture mock-up assembly, and

- the cigarette to be tested.

A photograph of a test chamber containing a mock-up assembly and a cigarette is shown in Figure 1. The test procedure is fully described in Appendix B. The following gives a brief description of the test method.

This test procedure begins with the operator preparing the mock-up assemblies in a conditioned environment. Clean, gloved hands are used at all times during the test procedure when handling mock-ups and cigarettes (to preclude salt contamination). The mock-ups and cigarettes are conditioned for at least 24 hours at 55 ± 5 % relative humidity (RH) and 23 ± 3 °C. After conditioning, the test materials may be moved from the conditioning room/chamber to the test room in sealed plastic bags just prior to testing. The test room is conditioned to the same relative humidity and temperature levels as the conditioning room. (Note the test room conditioning was specified somewhat differently in the preliminary interlaboratory study). The vacuum draw ignition apparatus is calibrated to a flow of 1000 cc/min. The mock-up assembly is placed into the test chamber's center and a cigarette test specimen is selected and weighed. If the cigarette weight falls within the required test range for that lot of specimens, a pencil mark is placed on the seam side, 15 mm from the tip. The vacuum draw apparatus is started and the cigarette is placed into the apparatus holder. A butane gas cigarette lighter with a pre-set, 15 mm high flame is ignited and held to the end of the cigarette for three seconds. The lit cigarette is carefully removed from the ignition apparatus and is moved to the test chamber where it is placed into a cigarette holder located on the center of the mock-up assembly. The chamber door is closed, and the cigarette is allowed to burn down to the 15 mm mark. At this point, the cigarette and holder are removed from the mock-up. The cigarette holder is placed into the test chamber's corner and the cigarette is carefully placed diagonally across the mock-up assembly with the ash located at the center of the mock-up. A stopwatch is started to measure the burning time of the cigarette. If the ash falls off at any point in this process, another cigarette is selected; and the process starts again as above. The cigarette is allowed to burn until one of the following occurs:

- self-extinction of the cigarette,

- the cigarette burns its entire length without igniting the mock-up assembly, or

- ignition of the mock-up assembly.

An ignition is defined as a char zone propagating away from the burning tobacco column by at least 10 mm. The stopwatch is stopped upon observing any of the three final test conditions described above. If the mock-up ignites, it and the cigarette are carefully extinguished. The test results are recorded.

8. Interlaboratory Study of Mock-Up Method

a. Preliminary Considerations

All test methods have some random variation that cannot be controlled easily. Tests performed on materials considered to be identical under presumed identical test conditions do not, in general, produce identical test results. This random behavior is generally attributed to the operator, equipment used, calibration of the equipment and environmental changes. Controllable variability is kept to a minimum by a good written test procedure.

Standardized techniques have been developed for the evaluation of test method variability and precision. Precision, as defined by ASTM, is a concept related to closeness of agreement among test results obtained under prescribed like conditions from a measurement process being evaluated [24]. The approach used to evaluate the precision of a test procedure is an interlaboratory study (ILS), referred to also as a round robin. The guide used for planning the interlaboratory studies reported in this report was ASTM E691, Standard Practice for Conducting an Interlaboratory Study to Determine the Precision of a Test Method [25].

Results from an interlaboratory study generally provide information on *repeatability*, i.e., a measure of variability within a laboratory, and *reproducibility*, i.e., a measure of variability between laboratories. In addition, interlaboratory studies are often used in the process of test method development since a properly designed experimental plan can help to identify areas of variability which may require additional control. In the work reported here, interlaboratory studies were used for improving the test procedures as well as for evaluating precision and reproducibility.

In planning the interlaboratory test programs reported here many factors were considered. Certain of these were viewed as vitally important. Each of these key requirements was taken from ASTM E691:

- A properly designed ILS will be as simple as possible in order to obtain estimates of within- and between-laboratory variability that are free of unnecessary interferences.

- The design should include at least six laboratories.

- Laboratories participating in an ILS must be qualified to conduct the test procedure.

Figure 1. Photograph of a test chamber containing a mock-up assembly and a cigarette.

- The test method should be subjected to a ruggedness test prior to being used in a major ILS. A ruggedness test is generally a small ILS which uses two or more laboratories for evaluating and adjusting requirements in the test method to enhance its function and to identify areas of variability which may need improvement.

- No fewer than three materials, in this case cigarettes, should be used in designing an ILS, and the materials should represent different levels of property measurement.

- The numbers of tests in an ILS should be of sufficient number to obtain a good estimate of repeatability.

b. Selection of Cigarettes for Interlaboratory Study

As described above (Section II.A), the cigarettes for the round robin studies were selected from the Series 500 set; there were insufficient cigarettes from the Series 100 set for this purpose. Series 500, like Series 100, includes 32 different cigarette designs (*i.e.*, variants of tobacco type, packing density, paper citrate content and paper porosity); a smaller subset was chosen for use in the interlaboratory studies.

The size of the subset to be used in the ILS clearly affects the total testing load to be imposed on all participating laboratories, and a compromise between cigarette design diversity and test load was sought. These concerns led to the choice of a three-week test program for the mock-up ignition method and a one-week plan for the cigarette extinction method. The experimental plans were designed to use a balanced selection of five different cigarette types for the study.

Eight of the thirty-two cigarettes in the 500 Series were initially selected as candidates to be used in the ILS. (See Sect. II.A.) These initial cigarette types were chosen to reflect the range of designs found in this group of experimental cigarettes. Packing parameters used in the selection included tobacco type, expanded vs. nonexpanded tobacco, wrapping paper porosity, paper citrate content and cigarette circumference. This initial selection of cigarettes consisted of types identified with the following numbers: 501, 503, 506, 508, 520, 529, 530 and 531.

The second phase of selection, which picked the five cigarettes to be used in the interlaboratory study, was based on the range of ignition performance. Tests were conducted to identify the ignition propensity of the eight cigarettes using the three mock-up assemblies selected for the interlaboratory study. The results are shown in Table 5 (Section II.A). On the basis of these results, the following cigarette types were chosen for the interlaboratory study: 501, 503, 529, 530 and 531. Cigarettes 501 and 503 have relatively high ignition propensities; cigarette 531 has an intermediate ignition propensity and cigarettes 529 and 530, relatively low propensities. The choice of these five cigarettes provides a range of performance which can be used to evaluate the test procedure appropriately. This range of ignition propensity covers that from the population of the experimental cigarettes supplied by the industry for this study. Prior NIST work [3] has shown that the high end of this range was typical of current commercial cigarettes, while the lower end tends to cause few ignitions on any of the tested substrates. Table 19 provides a description of each cigarette type used in the interlaboratory study.

Table 19. Description of Interlaboratory Study Cigarettes

Cigarette Designation	Tobacco Type	Tobacco Expansion	Paper Porosity	Paper Additive	Circumference (mm)
501 BNLC-21	Burley	Non-Expanded	Low	Citrate	21
503 BNHC-21	Burley	Non-Expanded	High	Citrate	21
529 FELC-25	Flue-Cured	Expanded	Low	Citrate	25
530 FELN-25	Flue-Cured	Expanded	Low	None	25
531 FEHC-25	Flue-Cured	Expanded	High	Citrate	25

c. Logging and Randomizing Mock-Up Materials

<u>Logging of Samples.</u> Several systems were implemented to track mock-up materials from product lots. Log books were maintained for the receipt and identification of all materials. Similarly, records were kept on all materials sent to the individual laboratories participating in the interlaboratory study.

Fabric bolts were prepared in runs of approximately 64 linear meters (70 linear yards), and the bolts were numbered sequentially. Each bolt was given an identification number. When a bolt was selected for cutting, the fabric was laid out and marked into 20.3 cm x 20.3 cm (8" x 8") samples. Every sample was identified with a duck number, a bolt number and two additional numbers which indicated the length and width position of the sample in the individual bolt. All numbers identifying a test sample were entered into a permanent log book. The fabrics were handled by gloved personnel and maintained in closed plastic bags prior to mock-up preparation. At approximately 10 meter intervals, a sample was randomly selected from across the width of the goods for ion chromatography analysis.

The polyethylene film samples were tracked, prepared and identified in the same manner as was used for the fabrics. At approximately every 3 linear meters (10 linear ft), a sample was taken for product analysis testing.

The polyurethane foam order consisted of three buns from a sequential production lot. NIST sent an observer to the production plant to verify how the foam was formed, cured, cut and packaged. The lots were marked to indicate the orientation of the foam as it was received off the production run. The packages were disassembled at NIST, and individual foam samples from two of the buns were identified by length and width from the section of the production lot. Every foam piece was logged into a permanent record book. The foam was maintained in closed cardboard boxes.

<u>Randomization of Samples.</u> Fabric samples were randomized according to the following procedure. First, the total number of a cotton duck fabric samples (*e.g.*, duck #4) needed for testing throughout nine laboratories was determined. That number was apportioned, for nearly even distribution, from the number of possible samples obtainable from each bolt of that duck. The appropriate number of samples from a given bolt, was taken randomly and then distributed randomly among the nine laboratories. Laboratory sample logs were prepared by NIST to track samples being sent to the labs

(duck no., bolt no., length and width position of sample on the bolt). Test laboratories were instructed to randomize the samples for any given fabric type.

The polyethylene film samples were randomized in the same manner as the fabrics.

Eleven subsections of polyurethane foam were selected at random from the production lot. The individual samples from the foam subsection were identified with two symbols. The foam pieces were then randomly distributed throughout a large, clean room. A number of NIST staff members were asked to randomly select five pieces of foam from the room and place the foam into cardboard boxes. This was then repeated in turn for each foam subsection. At the end of this process, each box contained 55 pieces, 5 pieces from each of the 11 subsections. The test laboratories were instructed to take one box of foam for a given day's testing and to randomize those foam pieces prior to testing.

d. Preliminary Interlaboratory Study

A preliminary interlaboratory study was conducted for evaluation and further refinement of the mock-up ignition method. This study was not designed to validate the new procedure but rather was designed as a screening round to evaluate the effectiveness of the written test protocol and to further study the test method on a multi-laboratory basis. This preliminary round also met the need for a ruggedness test prior to conducting a complete ILS. Three laboratories participated in the preliminary study: Consumer Product Safety Commission, Engineering Laboratory; National Institute of Standards and Technology, Building and Fire Research Laboratory; and Philip Morris USA, Research Laboratory.

In June, 1992, a memo was sent to each participating laboratory providing basic information about the planned study. This memo included a draft of the test method and identified areas where the laboratories might have to make modifications to their test facilities needed for successfully conducting the study. Emphasis was placed on the need for tight control over environmental conditions in the specimen conditioning room/chamber and in the test room. The requirements called for the conditioning room/chamber to be maintained at 50 ± 5% relative humidity (RH) and 23 ± 3 °C and the test room to be maintained at 55 ± 10% RH and 23 ± 3 °C.

<u>Test Operator Training.</u> Each laboratory sent two test operators to NIST for training in July, 1992. One trainee was to be experienced with cigarette ignition testing and the other was to possess only general laboratory skills with no fire test experience. This difference in operator skills would be one of the variables in the ILS. During this training session operators also received detailed instructions on how to report test results. All test operators received a test workbook which contained a copy of the test procedure, a daily weather information form, a test procedure checklist, a fifteen day experimental plan and a daily experimental plan specific to each operator. This book also contained a sample, filled-in worksheet as a guide for the operators and a set of blank individual test worksheets for reporting all tests. In addition, each laboratory received a computer disk containing a program for entering their daily test results. The computer data were used as a backup for the workbooks and also facilitated preparation of a computer-readable data base for use in the data analysis.

<u>Test Chambers and Accessories.</u> Test chamber kits with square brass frames for holding the fabric/film flat on the foam substrate and cigarette holders were prepared at NIST. Several weeks before testing was to begin, the test chamber kits with all accessories and two butane cigarette lighters

were shipped to each of the laboratories. The chamber kits provided enough materials to construct five complete test chambers, although only four were needed for the study. Each laboratory assembled their own chambers using directions supplied with the kits.

Test Materials: Cigarettes. Before shipping test cigarettes to the laboratories, NIST took a random sample of cigarettes from each lot and weighed them to determine the acceptable weight range for cigarettes to be tested. The test weight range was plus or minus two standard deviations from the mean value of the sample. A weight range table was prepared and sent to each laboratory with the cigarettes. The participants were instructed to use only cigarettes that fell within the weight ranges specified in the table. All cigarettes that exhibited weights outside of the specified ranges were to be discarded.

Cigarettes were randomly selected from each lot for each laboratory and packaged for shipping. Approximately 200 cigarettes of each test type were shipped to the laboratories by two-day delivery. This quantity provided enough cigarettes to allow for losses resulting from specimens that were out of the acceptable weight range or were damaged and for retests if materials were discarded from aborted tests.

Test Materials: Mock-Ups. The three mock-up assemblies described earlier in the text were used: duck #4 with a layer of polyethylene film placed between the fabric and polyurethane foam, duck #6 placed directly atop the polyurethane foam, and duck #10 placed directly atop the polyurethane foam. The experimental plan required each substrate to be tested with each cigarette type 48 times (24 times by each operator). Approximately 280 sets of fabric and foam for each type of mock-up were randomly selected for each laboratory and shipped to them for testing. This provided approximately forty extra mock-up assemblies for each type used in the study. The excess assemblies allowed the laboratories to replace damaged materials or rerun aborted tests.

Laboratory Visits. During the month of August, 1992, the ILS coordinator visited each of the participating laboratories. These visits included a review of laboratory arrangements for testing, an air flow calibration check for each test chamber, a standard relative humidity calibration for each laboratory, and a review of the test program protocol and test method. The visit also provided opportunities for discussing any last minute questions which the participants had before beginning the test program. The preliminary test program began during the last week of August; and all laboratories had completed the test program by the end of September, 1992.

Nature of the Preliminary Test Round. This preliminary test program was carried out using the mock-up ignition method described above. The interlaboratory test plan was developed with assistance from the NIST Statistical Engineering Division, using ASTM E691-87 [25] as a guide. The factorial design used had the following structure:

- 3   Laboratories
- 2   Operators per laboratory
- 5   Cigarette types
- 3   Number of substrates
- 4   Number of test chambers
- 48   Replicates per cigarette per mock-up
- 3   Weeks of testing
- 720 Total cigarette tests per laboratory

Within the factorial experimental design, the following variables were tracked for possible study:

- Operator skill level - experienced or unexperienced
- Time of day - morning (AM) or afternoon (PM)
- Test chamber number - 1, 2, 3 or 4
- Mock-up assembly type - 1, 2 or 3
- Conditioning room relative humidity and temperature
- Test room relative humidity and temperature
- Cigarette ignition propensity

<u>General Test Plan.</u> All tests were to be performed in the prescribed randomized order as specified in the individual operator workbooks. A single cigarette type was tested by both operators on any given day. Both operators conducted their specified tests simultaneously. Each operator was assigned a pair of test chambers to be used during the morning hours and then switched to their co-worker's test chambers during the afternoon. Mock-up assemblies were tested in the order specified in each operator's workbook. The plan resulted in each cigarette/mock-up assembly being tested twice on each day. Individual test results were to be recorded in the workbooks as each test was completed; and each operator was required to complete a daily summary sheet containing all the information on laboratory operations, conditioning room/chamber control and environmental control in the test room. At the end of each day, operators were requested to transfer their data from the workbook to the computer disk data file.

<u>Analysis of Results.</u> When the test workbooks and computer disks were received at NIST, each was carefully reviewed for accuracy. A small percentage of errors of various types was found in the booklets and computer files. The workbooks showed some missing data and showed some mixed units, generally in temperature measurements. The computer files exhibited typos, transposed numbers and mixed units. These irregularities were corrected on the computer files (by reference to the workbooks) before the data were transferred onto combined laboratory computer files and submitted to the NIST Statistical Engineering Division for analysis.

The combined data file contained 2160 (= 720 × 3) lines of data, corresponding to 720 ignition tests per lab for each of 3 labs. Each line of data consisted of the values of 13 variables. The names used for these variables and a description of the information they represent are summarized in Table 20.

Table 20. Variables in Analysis of Preliminary Interlaboratory Study

Variable Name	Description
TST_RSLT	Test Result, coded as: I=Ignition, N=Non-Ignition, S=Self-Extinguishment
LAB	Laboratory Number (1-3)
CIG_TYPE	Cigarette Type (Coded as 1-5, representing Series 501, 503, 529, 530, 531, respectively)
SUBSTRAT	Fabric/Film/Foam Substrate Identifier 1 = Number 4 Cotton Duck 2 = Number 6 Cotton Duck 3 = Number 10 Cotton Duck
Auxiliary Categorical Variables:	
CHAMBER	Test Chamber Number (1-4)
TST_BLK	Test Block (Week of testing, or equivalent group of five test days = 1, 2 or 3)
OPERATOR	Operator (E=Experienced, I=Inexperienced)
AMPM	Time of Day (A=AM, P=PM)
DATE	Date of test (MMDDYY)
Auxiliary Continuous Variables:	
TSTTEMP	Test Room Temperature
TSTRH	Test Room Relative Humidity
CNDTEMP	Conditioning Room Temperature
CNDRH	Conditioning Room Relative Humidity

Except for DATE, all of the variables in Table 20 were studied in the statistical analyses. The DATE variable was used primarily in the process of checking the data files.

In reporting the test results (TST_RSLT), the laboratories made a distinction between two distinct types of non-ignition outcomes, as follows. For a cigarette which extinguished before the entire tobacco column was burned, the outcome was coded as S, for Self-Extinguishment. Alternatively, when the tobacco column burned all the way to the end without igniting the fabric substrate, it was coded as N, for Non-Ignition.

The test results for the preliminary round are summarized by LAB, CIG_TYPE, and SUBSTRAT in Table 21. The distinction shown there between Self-Extinguishment and Non-Ignition was not used formally in the statistical analysis of the results. Instead, a simpler presentation and analysis were obtained by combining the two types of non-ignition. Thus, a derived variable, named "IGN," was defined as follows:

IGN = Y if TST_RSLT = I (ignition)
= N if TST_RSLT = N or S (non-ignition).

A graphical summary of the test results for the preliminary interlaboratory study, based on the derived variable, IGN, is shown in Figure 2. In this figure, the height of each vertical bar represents the proportion of test runs resulting in ignition (IGN=Y) obtained by the corresponding laboratory for the substrate and cigarette indicated. The 15 bar charts are arranged in a pattern with three rows, corresponding to the three substrates (mock-up configurations) used in testing, and five columns corresponding to the five cigarette types tested. The order in which the cigarettes are shown is based on the total number of ignitions for each cigarette type, with cigarettes having the highest ignition propensity on the left and those having the lowest ignition propensity on the right. (Cigarettes 503 and 501 actually had the same number of ignitions in the preliminary interlaboratory study. Cigarette 503 is shown first in Figure 2 based on the fact that 503 had the most ignitions in the main interlaboratory study described below. Except for the tie between cigarettes 503 and 501 in the preliminary round, the ordering of the cigarettes based on total number of ignitions was the same in the two rounds of interlaboratory tests of the mockup ignition test method.)

It should be observed from the summary shown in Figure 2 and Table 21 that the lab-to-lab variation in the proportion of ignitions is not excessive in comparison with the amount of variation that is commonly found in fire testing. (See below in discussion of main interlaboratory study.) In fact, the largest deviation of any single lab value from the mean proportion of ignitions was about 0.15, which occurred for cigarette 529 on substrate 3 (duck #10). Thus, based on this simple criterion, the mockup ignition test method showed promise of utility.

The participating laboratories were instructed to control the test environment so as to maintain the temperature and humidity variables within defined limits. The data showing the actual range of these variables in the preliminary round are summarized graphically in Figure 3.

Table 21. Summary of Test Results for Preliminary Interlaboratory Study

Cigarette Type	Substrate	Laboratory	Test Results		
			Ignitions	Non-Ignitions	Self-Extinguishments
1	1	1	48	0	0
		2	40	8	0
		3	36	12	0
	2	1	48	0	0
		2	48	0	0
		3	48	0	0
	3	1	48	0	0
		2	48	0	0
		3	48	0	0
2	1	1	47	1	0
		2	35	13	0
		3	42	6	0
	2	1	48	0	0
		2	48	0	0
		3	48	0	0
	3	1	48	0	0
		2	48	0	0
		3	48	0	0
3	1	1	0	0	48
		2	0	0	48
		3	0	0	48
	2	1	3	0	45
		2	0	0	48
		3	3	0	45
	3	1	13	0	35
		2	8	0	40
		3	21	0	27

Cigarette Type	Substrate	Laboratory	Test Results		
			Ignitions	Non-Ignitions	Self-Extinguishments
4	1	1	0	0	48
		2	0	0	48
		3	0	0	48
	2	1	0	0	48
		2	0	0	48
		3	0	0	48
	3	1	0	0	48
		2	0	0	48
		3	1	0	47
5	1	1	4	23	21
		2	0	17	31
		3	3	34	11
	2	1	45	1	2
		2	37	0	11
		3	46	0	2
	3	1	47	0	1
		2	45	0	3
		3	47	0	1

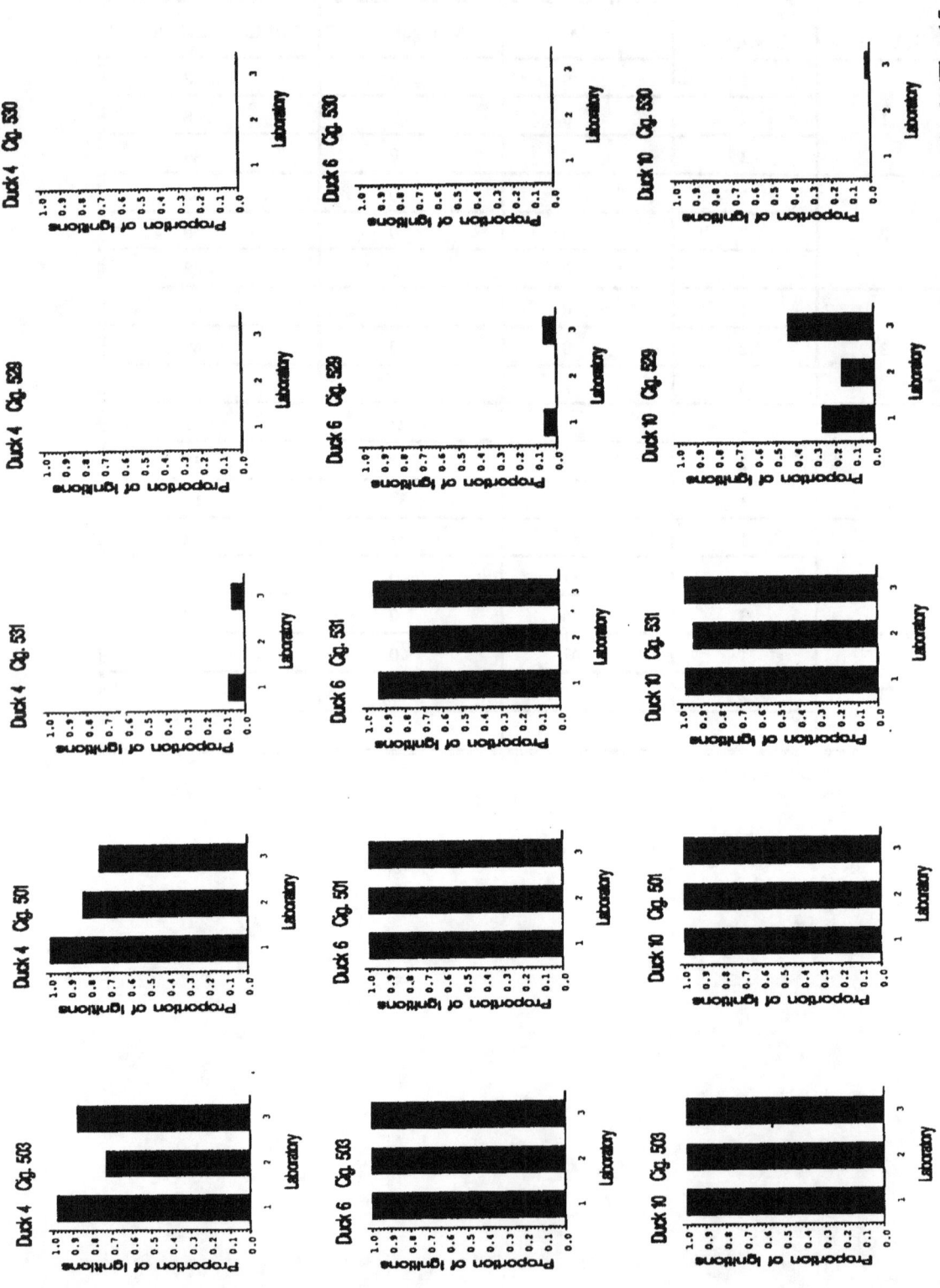

Figure 2. Comparison of Ignition Rates for the Preliminary Interlaboratory Study of the Mock-up Ignition Test Method. The 15 plots in the figure correspond to the 5 cigarette types tested, by columns, and the 3 test substrates, by rows. In each component plot, the vertical bars represent the proportions of ignitions obtained by each of the 3 participating laboratories.

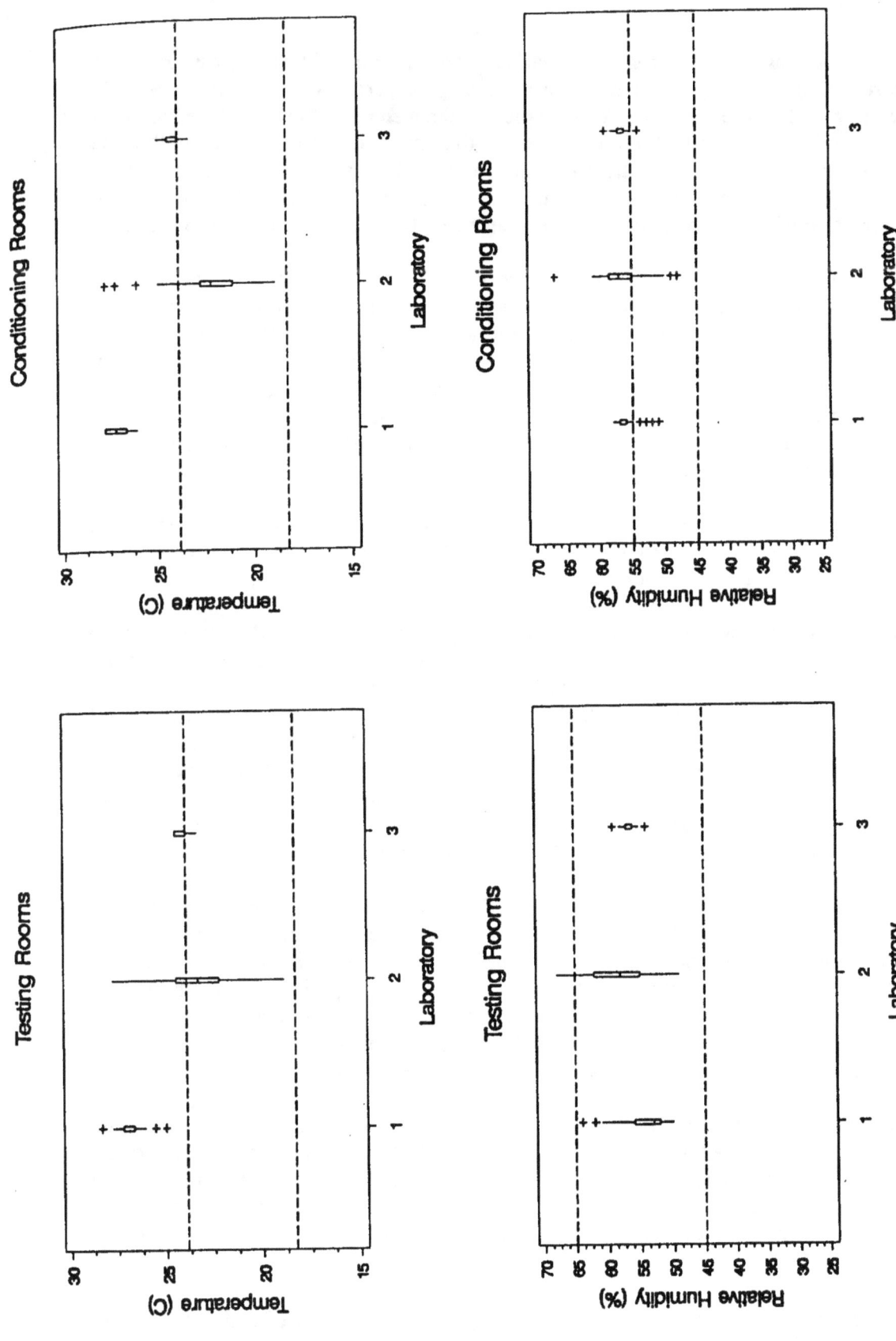

Figure 3. Environmental Conditions Reported by the Laboratories Participating in the Preliminary Interlaboratory Study of the Mock-up Ignition Test Method.

In Figure 3, the distributions of the environmental variables are represented by box plots. Box plots are constructed so that the rectangular boxes contain the range covered by the central 50% of the data, with a line drawn in the interior of the box to represent the median of the data. The "whiskers" (vertical lines) attached to the central 50% boxes extend to the upper and lower limits of all the data, except that values far away from the central portion are identified as "outliers" and are plotted separately with "+" symbols. (The statistical software used to produce these box plots identifies an outlier as any data value whose distance from the median exceeds 1.5 times the length of the box.)

Also indicated in Figure 3, by the space between the horizontal dashed lines, are the target ranges for the environmental variables stated in the instructions for conducting the testing. Comparing the locations of the box plots with the dashed lines shows how successful, or not, the labs were in maintaining the environmental conditions within the target ranges. For example, the plot in the upper left-hand corner of the figure shows clearly that temperature in the testing room of Lab 1 was always a bit higher than the target range. The same plot shows that, while the temperature varied more widely in Lab 2 than in Lab 1, the temperature in Lab 2 was inside the target temperature range approximately two-thirds of the time.

For all three labs, the humidity in the conditioning rooms tended to stay just above 55%, which was the upper limit of the target range. As mentioned in Section II.B.8.e below, this fact influenced the decision to change the target range for humidity in the main interlaboratory study.

A statistical analysis was undertaken to ascertain whether a significant portion of the variation in the ignition results could be attributed to variation in the four environmental variables: TSTTEMP, TSTRH, CNDTEMP, and CNDRH. The statistical procedure used was logistic regression analysis, implemented by the CATMOD procedure in SAS®.[6] This analysis proceeded by first fitting a full model in which IGN was modeled as a function of the variables LAB, CIG_TYPE and SUBSTRAT, plus the four continuous environmental variables. This model was compared with a reduced model in which the environmental variables were omitted. The overall result of this analysis was that *none* of the environmental variables showed a statistically significant effect on ignitions, based on the criterion that the significance probability, or "p-value," was greater than 0.05.

By contrast, the other three variables, LAB, CIG_TYPE and SUBSTRAT were all statistically significant well below the 0.05 significance level. This result was not unexpected. The five cigarettes were selected for testing to represent a range of ignition behaviors. Similarly, the three substrates were designed to elicit a range of ignition responses. The finding of significant variation between labs, while not particularly desired, is nonetheless a common occurrence in interlaboratory studies of test methods of all kinds. General experience with interlaboratory studies at NIST has been that a statistically significant difference between laboratories can be detected more often than not.

The statistical analysis also investigated whether the ignition results were significantly affected by the four categorical variables: CHAMBER, TST_BLK, OPERATOR, and AMPM. Both logistic regression and the Cochran-Mantel-Haenszel test procedure were used for this purpose, again implemented by the CATMOD procedure in SAS®. The Cochran-Mantel-Haenszel procedure was implemented in a way which tests for a significant association between ignition results (IGN) and each

---

[6] SAS® statistical software, Version 6. SAS® is a registered trademark of SAS Institute Inc., SAS Circle, Box 8000, Cary, NC 27512-8000.

categorical variable while controlling for the effects of LAB, CIG_TYPE and SUBSTRAT. For example, when applied to the OPERATOR variable, these procedures test whether the difference in ignition results between the experienced and inexperienced operators was greater than would be expected due to random variation. As was the case for the environmental variables, these tests were all nonsignificant, using a significance level criterion of 0.05.

Results from the preliminary ILS showed that the test procedure could be successfully replicated in more than one laboratory. The results from the three laboratories also indicated that the test procedure could be expected to be acceptably repeatable within a laboratory and reproducible between laboratories. The findings from this study pointed the way to the main ILS.

e. Main Interlaboratory Study

The main interlaboratory study was similar to the preliminary version, but a number of changes were made based on findings from the earlier study. The same factorial experimental design was used, expanded to include nine laboratories.

With this test round, as before, each new laboratory was shipped test chamber kits which they assembled. With this shipment they also received the square brass frames and cigarette holders for each chamber. An operator training program was held in September, 1992 for all participating laboratories. Again, each laboratory was asked to provide an operator experienced with cigarette fire testing and an inexperienced tester. During the training session, each operator acquired hands-on experience with the mock-up ignition method and the cigarette extinction method. Both test methods were included in the training because an ILS for the cigarette extinction method was planned to follow the ILS for the mock-up ignition method.

Based on the satisfactory performance of the test cigarettes and mock-up assemblies in the preliminary round, each was retained for the main ILS. Each of the participating laboratories was shipped approximately 200 cigarettes of each type, as used in the preliminary study. The three sets of mock-up test materials that duplicated those used in the preliminary study were also shipped to each laboratory. However, the original lot of polyethylene film had been depleted. Additional polyethylene film was ordered from the retail supplier, and they inadvertently substituted a nominally similar material, which was included in the shipments to the nine laboratories. The film difference was discussed in Section II.B.3.

An additional 125 cigarettes with a separate designation ("F") were sent to the labs with the other cigarette lots and an extra box of fabric and foam substrate materials was shipped to each laboratory. These additional cigarettes and mock-up materials were provided to give the test operators an opportunity to practice running the test before beginning the main study. The practice tests would only be run after a visit by the interlaboratory study coordinator. Test operators were instructed to use cigarette "F" and the extra box of fabric and foam materials to run twelve practice tests. After completing the practice tests with cigarette "F," each lab would report their results to NIST by FAX.

As was done with the preliminary ILS, the interlaboratory coordinator visited each laboratory to review laboratory environmental conditions and control, check test chamber functioning and provide a relative humidity calibration for the lab. Also the coordinator met with laboratory management and test operators to review the test protocol and to answer last minute questions. All laboratory visits were completed by the middle of October, 1992. Most laboratories began testing the practice cigarette materials shortly after the coordinator visit was completed. Upon receiving the reports by FAX on the practice round, NIST reviewed the data; and laboratories were given permission to start the main round. All laboratories completed their assigned test work and submitted their data to NIST by the middle of December, 1992.

Impact of Preliminary Round: Environmental Conditions. In the preliminary study, the conditioning room and test room environments had different specifications. Conditioning room/chamber relative humidity was to be controlled at 50 ± 5 %. The test room was to be controlled at a level of 55 ± 10% RH. This difference in environmental conditions appeared to create some confusion within the participating laboratories. Therefore, discussions with the laboratories and a review of test results brought about agreement on a change to the required conditions. When the test procedure was rewritten for the main ILS, it included a single new environmental requirement for both conditioning room and test rooms: 55 ± 5% RH. This modification of the procedure also allowed the cigarette company labs to operate under conditions normal to their needs, and it also allowed laboratories using test procedures requiring relative humidities of 50% to maintain their normal laboratory conditions.

Impact of Preliminary Round: Selecting Cigarette Test Weights. In the preliminary ILS, NIST sampled and weighed each test cigarette lot to establish the range of cigarette weights to be tested. In the main study, this process was transferred to each laboratory. The laboratories were shipped approximately 200 cigarettes of each type and were instructed to follow this same procedure in determining the proper weights of cigarettes to be tested from each lot.

Impact of Preliminary Round: Cigarette Self-Extinction. In the preliminary ILS, it was found that certain types of cigarettes had burning characteristics that resulted in self-extinction shortly after they were ignited. Such a cigarette would often go out before being laid onto the mock-up assembly; hence it was discarded and the test was re-run. Even if one of these cigarettes burned long enough to be placed onto the mock-up, it would soon self-extinguish. It became apparent that this portion of the test procedure could be modified to indicate a self-extinction at any point after a cigarette was properly ignited. During the main ILS, it was specified that if a cigarette should self-extinguish at any point after being properly ignited, even if it had not been placed onto the mock-up, the test was complete. Data from such cases were recorded as self-extinctions.

Participants and Procedures. Nine laboratories participated in the main interlaboratory study. This group included industry, state and federal government laboratories and an independent testing laboratory. The list of participants follows:

- American Tobacco Company, Research Laboratory
- Brown & Williamson Tobacco Company, Research Laboratory
- Bureau of Home Furnishings, State of California
- Consumer Product Safety Commission, Engineering Laboratory
- Diversified Testing Company
- Lorillard, Research Laboratory
- National Institute of Standards and Technology, Building and Fire Research Laboratory
- Philip Morris USA, Research Laboratory
- R.J. Reynolds Tobacco Company, Research Laboratory

[Note that this listing does not correspond to the identification numbers, 1 through 9, which appear later in the report. Those numbers were assigned randomly for anonymous presentation of the results.]

As stated above, several changes were made to the test procedure based on experiences gained from the preliminary ILS. The preliminary study also indicated how the recording and reporting of data could be streamlined, and these changes were made in the main study. Each tester in the program received a booklet containing copies of the test procedure and other ILS guidelines. They also received three separate workbooks for recording data, one for each week of the three-week test program. Each laboratory received three computer disks for putting the test results into computer-readable form. The following instructions were given on new procedures for recording and submitting data to NIST for analysis: Test operators were directed to record all data in their workbooks as they prepared and completed each test. At the end of each day they were to enter their data into the computer, which would then print out a daily summary for each operator. These summary sheets were to be FAXed to NIST at the end of each day. At the end of each week, the weekly workbooks and the weekly computer disks were to be mailed to NIST by two-day delivery. This method of acquiring data helped to reduce the time needed for assembling the test data and preparing it for analysis.

Purpose and Methods of Analyzing Results. Information provided in the daily FAX report from the laboratories was used by NIST to evaluate laboratory progress. It also provided information necessary for preparing data files for each laboratory. As the ILS progressed, not all of the laboratories consistently submitted the daily FAX reports. Some labs would accumulate several days of testing before FAXing the reports to NIST. As the test results were submitted to NIST, they were organized into appropriate data files used for review of accuracy and for analysis. These files were compared to the data recorded in the test workbooks, and corrections were made as needed. Again with this ILS, there were missing data in the workbooks and incorrect units reported. The computer files again showed typos, transposed numbers and mixed units. There were 6,480 test results returned from the main ILS. Of these data, approximately five percent contained errors of the types discussed above. One operator in each of two laboratories failed to enter all the ignition data in the lab workbooks; however, all data were entered into the computer files. Less than 0.3 percent of the ignition data were missing from the workbooks.

A few cases were noted in which laboratories ran some of the test replications on the wrong mock-up configuration, compared to what was called-for in the test plan. The result was that the number of tests actually conducted differed from the intended number (48 replications) by ± 2, for some combinations of cigarette type and mock-up configuration. These cases are reflected in rows of Table 22 for which the total number of Test Results reported (Ignitions + Non-Ignitions + Self-Extinguishments) sums to 46 or 50, instead of 48. In all cases, the total number of tests performed by the laboratory was correct.

Raw Data. As noted, the combined computer file contains 6480 lines of data, corresponding to 720 ignition results per laboratory for each of 9 laboratories. The 720 results per laboratory arise from 48 tests of each of 5 cigarettes on each of 3 substrates ($720 = 48 \times 5 \times 3$).

The raw data on each line of the computer file represent the same set of thirteen variables described previously in Table 20. There are two minor changes in the data for these variables compared to their descriptions in Table 20: the LAB variable has a range from 1 to 9 in the main round, rather than 1 to 3, as in the preliminary round; and the variable OPERATOR no longer differentiates one of the operators as "Experienced" and the other as "Inexperienced," but rather codes the two operators simply as number "1" and number "2," in no particular order. This change was made because it was impractical, and seemed unnecessary, to recruit truly inexperienced operators at each of the nine laboratories for the main round evaluation. Generally, both operators at each lab had some previous experience in testing ignition propensities of cigarettes or had sufficient acclimation after the first several days of testing to be regarded as experienced.

A summary of the test results, by LAB, CIG_TYPE and SUBSTRAT, is presented in Table 22. Note that identifying numbers were assigned *independently* to the laboratories in the preliminary and in the main rounds of the interlaboratory evaluation so, for example, laboratory number 3 in the preliminary round is not the same as number 3 in the main round. In Table 22, the ignition results are recorded in three categories, as given by the variable TST_RSLT. As was done for the preliminary round, the two types of non-ignitions were combined into a single category for the statistical analyses. Thus, as before, the test results were analyzed in terms of the derived variable IGN, which simply records the results as ignition (IGN=Y) or non-ignition (IGN=N).

A graphical display of the data in Table 22 is shown in Figure 4, where, for each cigarette (by columns) and substrate (by rows) the proportion of ignitions is represented by a vertical bar for each laboratory. The cigarette types are shown from left to right in order of decreasing ignition propensity, with cigarette 503 having the most ignitions (in the left-most column of plots) and cigarette 530 having the fewest ignitions (in the right-most column). The three mockup configurations are shown as rows in the figure, with the least ignitable substrate (duck #4) as the top row and the most ignitable (duck #10) as the bottom row of the figure. For several cigarette and substrate combinations, all laboratories showed either 100% ignitions (charts near the lower left corner of the figure) or 0% ignitions (those near the upper right corner). Cases with intermediate ignition percentages fall near the diagonal (upper left to lower right) in the figure.

Table 22. Summary of Test Results for Main Interlaboratory Study, Mock-Up Ignition Method

Cigarette Type	Substrate	Laboratory	Test Results		
			Ignitions	Non-Ignitions	Self-Extinguishments
501	1	1	2	46	0
		2	5	43	0
		3	1	47	0
		4	3	45	0
		5	1	47	0
		6	3	45	0
		7	6	42	0
		8	16	34	0
		9	11	37	0
	2	1	48	0	0
		2	48	0	0
		3	48	0	0
		4	48	0	0
		5	48	0	0
		6	48	0	0
		7	48	0	0
		8	46	0	0
		9	48	0	0
	3	1	48	0	0
		2	48	0	0
		3	48	0	0
		4	48	0	0
		5	48	0	0
		6	48	0	0
		7	48	0	0
		8	48	0	0
		9	48	0	0

Cigarette Type	Substrate	Laboratory	Test Results		
			Ignitions	Non-Ignitions	Self-Extinguishments
503	1	1	20	28	0
		2	26	22	0
		3	19	29	0
		4	33	15	0
		5	22	26	0
		6	19	27	2
		7	21	27	0
		8	34	14	0
		9	36	12	0
	2	1	48	0	0
		2	48	0	0
		3	46	0	0
		4	48	0	0
		5	48	0	0
		6	48	0	0
		7	48	0	0
		8	46	0	0
		9	48	0	0
	3	1	48	0	0
		2	48	0	0
		3	50	0	0
		4	48	0	0
		5	48	0	0
		6	48	0	0
		7	48	0	0
		8	50	0	0
		9	48	0	0

Cigarette Type	Substrate	Laboratory	Test Results		
			Ignitions	Non-Ignitions	Self-Extinguishments
529	1	1	0	0	48
		2	0	0	48
		3	0	0	48
		4	0	0	48
		5	0	0	48
		6	0	0	48
		7	0	0	48
		8	0	0	48
		9	0	0	48
	2	1	1	0	47
		2	3	1	44
		3	6	0	42
		4	3	0	45
		5	0	0	48
		6	3	1	44
		7	0	0	48
		8	6	0	42
		9	11	0	37
	3	1	10	0	38
		2	15	0	33
		3	18	0	30
		4	13	0	35
		5	5	0	43
		6	14	0	34
		7	12	0	36
		8	20	0	28
		9	24	0	24

Cigarette Type	Substrate	Laboratory	Test Results		
			Ignitions	Non-Ignitions	Self-Extinguishments
530	1	1	0	0	48
		2	0	0	48
		3	0	0	48
		4	0	0	48
		5	0	0	48
		6	0	0	48
		7	0	0	48
		8	0	0	48
		9	0	0	48
	2	1	0	0	48
		2	0	0	48
		3	0	0	48
		4	0	0	48
		5	0	0	48
		6	0	0	48
		7	0	0	48
		8	0	0	48
		9	1	0	47
	3	1	0	0	48
		2	0	0	48
		3	1	0	47
		4	0	0	48
		5	0	0	48
		6	0	0	48
		7	1	0	47
		8	3	0	45
		9	6	0	42

Cigarette Type	Substrate	Laboratory	Test Results		
			Ignitions	Non-Ignitions	Self-Extinguishments
531	1	1	0	20	28
		2	0	22	26
		3	0	27	21
		4	0	28	20
		5	0	25	23
		6	0	30	18
		7	0	24	24
		8	0	28	20
		9	0	31	17
	2	1	44	0	4
		2	42	0	6
		3	48	0	0
		4	47	0	1
		5	46	0	2
		6	48	0	0
		7	46	0	2
		8	44	0	4
		9	45	0	3
	3	1	48	0	0
		2	46	0	2
		3	47	0	1
		4	47	0	1
		5	45	0	3
		6	48	0	0
		7	47	0	1
		8	48	0	0
		9	47	0	1

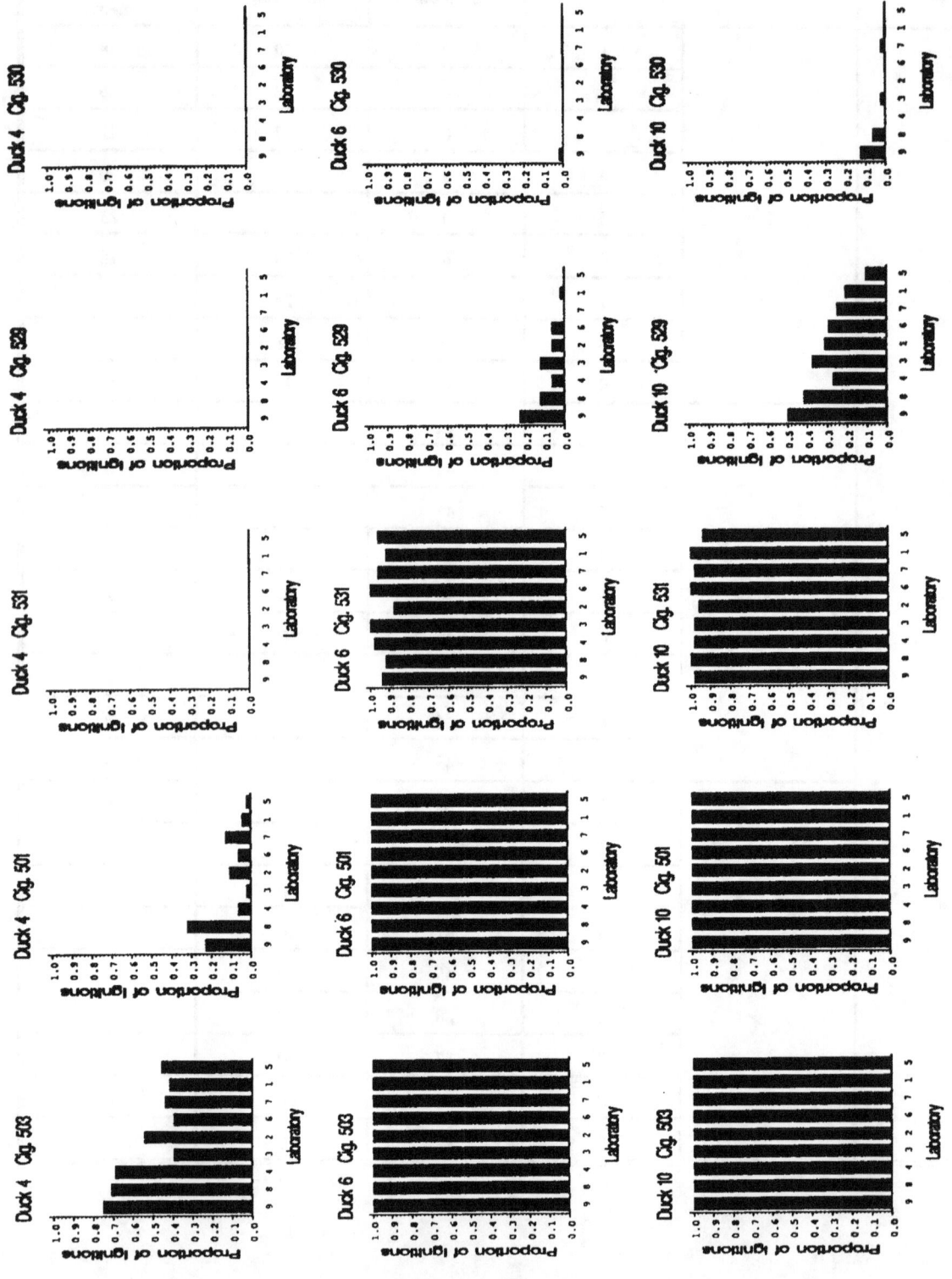

Figure 4. Comparison of Ignition Rates for the Main Interlaboratory Study of the Mock-up Ignition Test Method.

Auxiliary Variables. A statistical analysis was undertaken to detect possible dependence of the ignition results, through the variable IGN, on the eight auxiliary variables described in Table 20. It was not the purpose of this analysis to develop a detailed understanding of such dependencies from the interlaboratory study data — a designed experiment in a single laboratory would be better for that purpose. Rather, the purpose of the analysis of the auxiliary variables was to reveal features of the data that might point to any major problems with the execution or performance of the method in different laboratories.

The ranges of environmental conditions experienced by the laboratories are shown in Figure 5, using box plots to represent the range of data from each laboratory. Details of the interpretation of box plots are described later in this Section. The figure shows that most labs had at least a few temperature and/or humidity readings outside the target limits. Except for the temperatures in lab 8, the labs generally had most of their readings within the prescribed limits, or very nearly so.

Various statistical techniques were applied to investigate whether the ignition results were strongly influenced by the four environmental variables or by the discrete auxiliary variables, CHAMBER, TST_BLK, OPERATOR and AMPM. The Cochran-Mantel-Haenszel test was again used for the discrete variables. Conclusions from this test procedure showed no significant relation of IGN with CHAMBER or TST_BLK (*i.e.*, "week"), but did indicate the existence of possible effects due to both OPERATOR and AMPM.

In the case of OPERATOR, a statistically significant difference between the two operators was shown only for lab 4. In lab 4, the two operators obtained the same number of ignitions for 9 of the 15 cigarette/substrate combinations, but in each of the remaining six cases, the number of ignitions obtained by Operator 1 was always greater than the number obtained by Operator 2. The difference between the two operators was not large: Operator 1 obtained a total of just 12 more ignitions than Operator 2, out of 360 tests by each operator. It was mainly the consistency of the difference across the six cases that led to the attainment of statistical significance. In this study no attempt has been made to determine whether this observed difference can be traced to any differences in test procedure used by these two operators.

The variable AMPM, which indicates whether a test was run before or after noon, also showed a statistically significant, but small, effect. Overall, there was a slightly higher percentage of ignitions in the PM (47.4%) compared to the AM (45.7%). The Cochran-Mantel-Haenszel procedure, which identified this effect, is sensitive to both the size of the AM-PM difference and the consistency of its direction, while controlling for the LAB, CIG_TYPE and SUBSTRAT variables. While no satisfactory physical explanation was found for this overall effect, the following illustrates the kind of minor ambiguities that remain in the data. In a sub-analysis using data from lab 6 only, the increased ignition rate in the PM was also associated with an increase (of about 1 °C) in the average temperature of the test room (TSTTEMP). It is not possible with the information at hand, however, to establish a *causal* link between the two increases.

An attempt was made to perform a global analysis of the effects of all the study variables through the use of a detailed logistic regression model. This model-fitting exercise was not completely successful because of the large number of cells in the data matrix where the percentage of ignitions was either 0% or 100%. This feature of the data leads to a requirement of "infinite" parameter values in the model, with the additional result that the significance probabilities computed by the statistical software are not reliable, and therefore may not be worth pursuing.

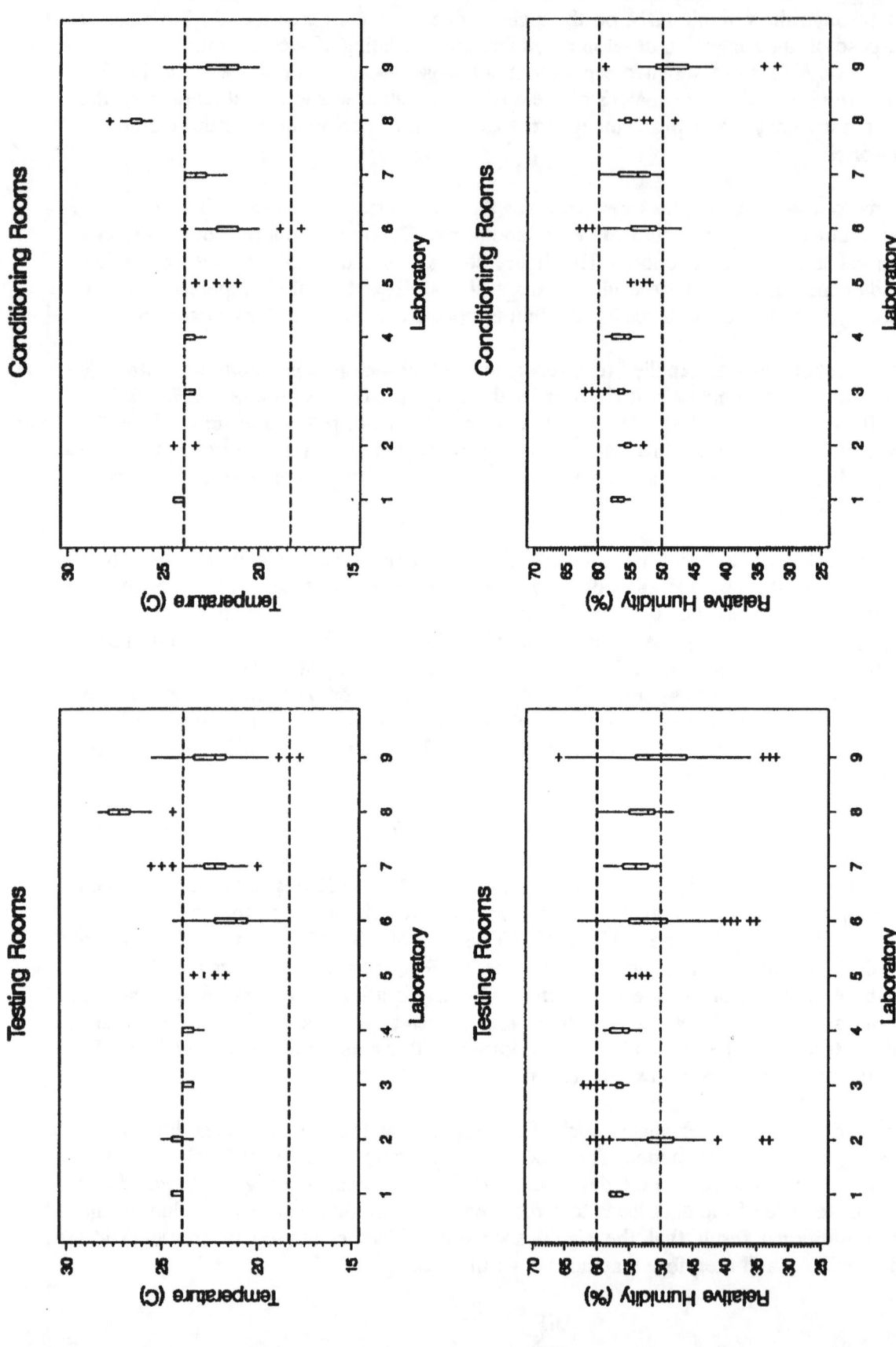

Figure 5. Environmental Conditions Reported by the Laboratories Participating in the Main Interlaboratory Study of the Mock-up Ignition Test Method.

A readily interpretable result from the global analysis pertains to the environmental variables. In the global model, the logarithm of the odds ratio in favor of ignition (IGN) was modeled as a function of all the variables (except DATE) in Table 20, together with several interactions. In this fitted model, the regression coefficients for all four environmental variables had the expected signs (though none had a statistically significant magnitude): the two temperature coefficients (for TSTTEMP and CNDTEMP) were positive, implying that increasing temperature increases the odds in favor of ignition, and the humidity coefficients (for TSTRH and CNDRH) were both negative, implying that increased humidity decreases the odds in favor of ignition.

Overall, the logistic analyses showed some indications of small, but possibly real, dependencies between the auxiliary variables and the ignition results. However, they did not reveal any major problems in the data. These indications are consistent with the general observation there will always be some means by which the test results could be improved by further refinements to the test method protocols.

Primary Variables. It is possible, and of some interest, to test whether the three primary explanatory variables in the data set, namely LAB, CIG_TYPE and SUBSTRAT, show statistically significant effects on the ignition results. In short, the IGN variables shows statistically significant effects due to all three of the primary explanatory variables.

For the LAB variable, statistically significant differences between labs were shown for essentially all cigarette/substrate combinations where the percentage of ignitions was far enough away from both 0% and 100% to allow differences to show. Specifically, significant between-laboratory differences were identified for cigarette types 503 and 501 on the duck #4 mock-up, for cigarette 529 on the duck #6 mock-up, and for cigarettes 529 and 530 on the duck #10 mock-up. These individual tests, for each cigarette and substrate, were combined into an overall test by the Cochran-Mantel-Haenszel procedure with the result that the overall hypothesis that there is no between-laboratory variation is rejected. The magnitude of the between-lab variability is summarized below through the estimation of repeatability and reproducibility standard deviations.

The significance test for differences between cigarette types was also carried out using the Cochran-Mantel-Haenszel procedure. The results showed that all five cigarette types have ignition rates that are statistically different from each other, with p-values less than 0.001 in each case. Turning to substrates, the Cochran-Mantel-Haenszel procedure also showed significant differences between the ignition results across all three mock-ups.

Repeatability and Reproducibility. In ASTM standard E-691, "Standard Practice for Conducting an Interlaboratory Study to Determine the Precision of a Test Method" [25], the summary precision statement for a test method under evaluation is based on statistical calculations of repeatability and reproducibility limits for the method. Recall that repeatability refers here to the consistency of test results from a single laboratory; reproducibility refers to the consistency of results among different laboratories. The calculation formulas described in ASTM E-691 are appropriate for test methods that yield measurements on a continuous scale, rather than the categorical "yes/no" outcomes that characterize the IGN variable of this study, and cigarette ignition testing generally. However, the repeatability and reproducibility measures can be adapted to categorical data applications, as will be described.

In ASTM E 691, the repeatability standard deviation is defined as the best estimate of the within-laboratory standard deviation of single measurement results. At the present stage of development, the definition of what constitutes a "single" measurement result for the Mockup Ignition Test Method has not been specified, except that it is understood that a single measurement result is the proportion of ignitions in some number (to be specified) of replications of the operation of placing a single type of lighted cigarette on a single type of mockup. For a cigarette having ignition rate P (a number between 0 and 1) on a given substrate, the standard deviation of the observed proportion of ignitions in m replications is equal to $[P(1-P)/m]^{1/2}$, based on standard properties of the binomial distribution. Some mathematical simplicity will be gained in what follows by initially taking m to be equal to the total number of replications actually conducted by each laboratory; $m = 48$ in the case of the mock-up ignition method. Further calculations using a simple statistical model will then allow comparisons of repeatability and reproducibility values that could be expected assuming different values of m.

Applying this approach to data for any substrate and cigarette type, the best (quasi-likelihood) estimate [26] of the *repeatability* standard deviation, S_r, is

$$S_r = \sqrt{\frac{\bar{p}(1-\bar{p})}{48}} \; ,$$

where $\bar{p}$ represents the mean proportion of ignitions across the nine laboratories and m has been set equal to 48. As defined, S_r is the best estimate of the pooled within-laboratory standard deviation, in that it combines information across all laboratories.

Using the convention that a single measurement result is defined as the proportion of ignitions in $m = 48$ replications, the number of single measurement results per laboratory is 1 in this study, and so the *reproducibility* standard deviation, S_R, as defined in ASTM E 691, is simply the between-laboratory standard deviation,

$$S_R = \sqrt{\frac{\sum_{i=1}^{9}(p_i - \bar{p})^2}{8}}$$

where p_i represents the proportion of ignitions for laboratory i and $i = 1, 2, \ldots 9$.

The repeatability and reproducibility standard deviations for the main round interlaboratory evaluation are summarized in Table 23.

Table 23. Observed Repeatability and Reproducibility Standard Deviations for Mock-Up Ignition Method
Main Interlaboratory Study; $m=48$ Replications per Laboratory

Cigarette I.D.	Substrate	Average Proportion of Ignitions	Repeatability S.D. S_r	Reproducibility S.D. S_R
501	1	0.110	0.045	0.102
	2	1.000	0	0
	3	1.000	0	0
503	1	0.532	0.072	0.145
	2	1.000	0	0
	3	1.000	0	0
529	1	0.000	0	0
	2	0.076	0.038	0.074
	3	0.303	0.066	0.117
530	1	0.000	0	0
	2	0.002	0.007	0.007
	3	0.025	0.023	0.043
531	1	0.000	0	0
	2	0.949	0.032	0.042
	3	0.979	0.021	0.021

As previously mentioned, the repeatability standard deviation is a function of the average proportion of ignitions, $\bar{p}$, for each cigarette type. In Table 23, it is apparent that the reproducibility standard deviation also depends on $\bar{p}$ to some extent. In order to provide a succinct summary of the data, it of interest to combine the estimates of repeatability and reproducibility across the substrates and cigarette types in this study.

ASTM Standard E 691 [25], in section 21.3, *Variation of Precision Statistics with Property Level*, gives the following general guidance in formulating precision statements for test methods where the repeatability and reproducibility depend on the level of the property being measured.

"Quite often the values of S_r and S_R will be found to vary with the values of the property level, $\bar{x}$ [corresponding to our $\bar{p}$].... The manner in which the statistics vary with the property level should be shown in presenting the precision information in the precision statement of the test method. The statistician should recommend the most appropriate relationship to present, using Practice E 177 as a guide."

ASTM E 177 [24], in section 28.5.4, recommends summarizing the relationship of the method precision to the property level "by a simple formula, or by a plot." For repeatability, the definition above specifies S_r as a simple function of the property level, $\bar{p}$. This relationship can be extended to accommodate the reproducibility standard deviation by use of a simple statistical model for what is variously called "extra binomial variation" [27] or "over dispersion" [26] or simply "heterogeneity" [28].

In this model, the between-laboratory variance, or reproducibility variance S_R^2, is related to the within-laboratory variance, or repeatability variance S_r^2, as follows:

$$S_R^2 = \frac{p(1-p)}{m}[1 + \varphi(m-1)]$$

$$= S_r^2[1 + \varphi(m-1)]$$

The quantity $[1 + \varphi(m-1)]$ in the above expression is called the "heterogeneity factor" in Finney's classic text on probit analysis [28]. The parameter φ in the heterogeneity factor plays the role of a correlation coefficient among replications performed within the same laboratory.

This model suggests that S_R^2 should vary roughly as a constant multiple of S_r^2. Figure 6 shows this approximate relationship for the data in Table 23, along with a least squares straight line through the origin given by

$$S_R^2 = (3.72)S_r^2.$$

Figure 6 shows that the simple statistical model gives a satisfactory summary of the relationship of S_R^2 to S_r^2 and, therefore, to $\bar{p}$. Identifying the slope of the least squares line in the Figure, 3.72, with the heterogeneity factor, $[1 + \varphi(48-1)]$, yields the estimated value $\varphi = 0.058$.

In terms of this model, then, the repeatability and reproducibility standard deviations can be approximated for any case of interest by using the formulas

$$S_r = \sqrt{\frac{p(1-p)}{m}}$$

$$S_R = \sqrt{\frac{p(1-p)}{m}[1+\varphi(m-1)]} \; ,$$

(1)

where p is the assumed ignition rate, m is the assumed number of replications per test result, and φ is the parameter for the heterogeneity factor, as estimated from relevant data ($\varphi = 0.058$ for the Mock-Up Ignition Test Method ILS.)

In Table 24, these formulas are used to calculate "repeatability limits," defined as $2.8 \times S_r$, and "reproducibility limits," defined as $2.8 \times S_R$, for the mock-up ignition method. The factor "2.8" in the definition of the repeatability and reproducibility limits is recommended in ASTM E 691 as a means to generate approximate 95% probability limits for the possible difference between two measurement results (*i.e.*, proportions based on m replications) obtained within the same laboratory (repeatability limit = $2.8 \times S_r$) or in different laboratories (reproducibility limit = $2.8 \times S_R$). For example, from Table 24 the reproducibility limit calculated as 0.39 for $m = 48$ runs means that, if the proportion of ignitions is obtained for $m = 48$ runs on the same cigarette/mock-up combination in each of two laboratories, then one might expect that the difference between the two proportions will be less than about 0.39 if the average cigarette ignition rate is near $p = 0.5$.

In interpreting the values in Table 24, the reader should bear in mind the statements in ASTM E 691 [25] that repeatability and reproducibility limits "should be considered as useful general guides," but "not exact mathematical quantities which are applicable to all circumstances and uses."

Table 24 allows one to compare the repeatability and reproducibility limits corresponding to several values of m, the assumed number of replications per "single measurement result." It is clear from the table, and from the formulas, that S_r is more strongly affected by increasing the number of replications than is S_R. This fact is highlighted in Table 24 by inclusion of the case where $m = 9600$ runs is assumed. In general, the repeatability decreases as the square root of m whereas the reproducibility approaches a non-zero limit for large m, which reflects the between-lab component of variability. This behavior shows the limitation to how much the reproducibility precision can be improved by increasing the number of replications within each laboratory.

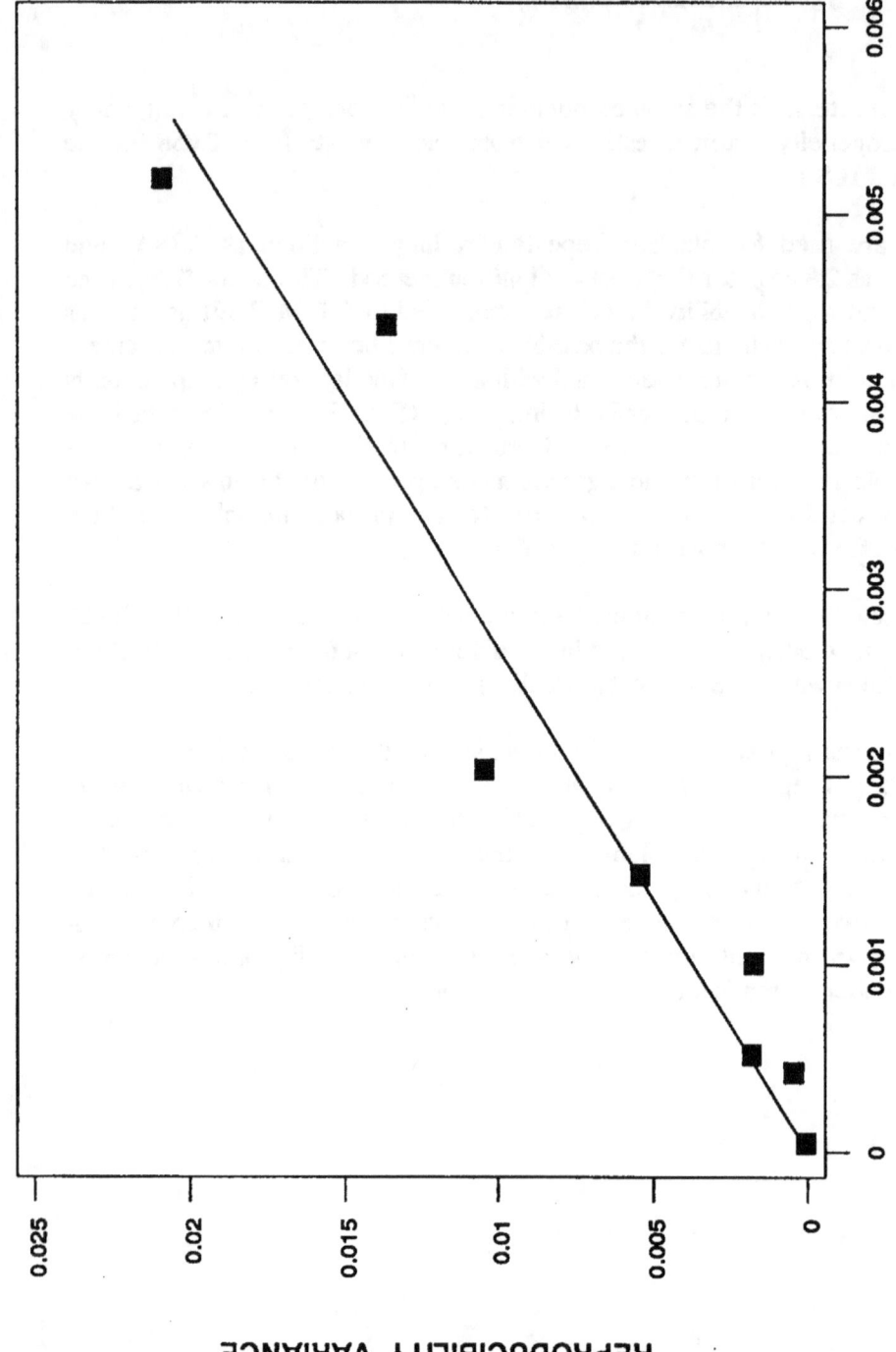

Figure 6. Empirical Relation of Reproducibility Variance, S_R^2, to Repeatability Variance, S_r^2, Based on Data in Table 23 from the Main Interlaboratory Study of the Mock-up Ignition Test Method. The equation of the least squares line shown is $S_R^2 = (3.72)S_r^2$.

**Table 24. Mock-Up Ignition Method:
Calculated Repeatability (r) and Reproducibility (R) Limits
for Various Assumed Numbers of Replications (m) and Ignition Propensities (p)**

p	m = 16		m = 32		m = 48		m = 96		m = 9600	
	r	R	r	R	r	R	r	R	r	R
0.05 or 0.95	0.15	0.21	0.11	0.18	0.09	0.17	0.06	0.16	0.006	0.15
0.10 or 0.90	0.21	0.29	0.15	0.25	0.12	0.23	0.09	0.22	0.009	0.20
0.20 or 0.80	0.28	0.38	0.20	0.33	0.16	0.31	0.11	0.29	0.011	0.27
0.30 or 0.70	0.32	0.44	0.23	0.38	0.19	0.36	0.13	0.33	0.013	0.31
0.40 or 0.60	0.34	0.47	0.24	0.41	0.20	0.38	0.14	0.36	0.014	0.33
0.50	0.35	0.48	0.25	0.41	0.20	0.39	0.14	0.36	0.014	0.34

m: Assumed number of replications per laboratory
p: Assumed long-run proportion of ignitions for cigarette and substrate combination under test
r: Repeatability limit = $2.8S_r$, where S_r is calculated from Equation (1)
R: Reproducibility limit = $2.8S_R$, where S_R is calculated as from Equation (1), with $\varphi = 0.058$.

Results from the main ILS show that the mock-up ignition test method can effectively differentiate the ignition propensities of various cigarettes, albeit at a limited degree of resolution. The most important limiting factor affecting the resolution of the test method is measured by the reproducibility limits. In Table 24, the repeatability limits (r) summarize the precision of the test method under the most favorable conditions for obtaining low variation (same lab, same equipment, short time period, same operators, etc.), whereas the reproducibility limits (R) measure the long-term stability of the test method.

The repeatability limits calculated in Table 24 represent the theoretical minimum amount of statistical variability that is inherent in data recorded as binary outcomes (ignitions vs non-ignitions). The amount by which the reproducibility exceeds the repeatability of this, or any, test method measures the degree to which unknown or uncontrolled influence factors affect the test results in the long term. Data from the nine laboratories in the ILS show that the ratio of repeatability to reproducibility limits is $R/r = S_R/S_r = \sqrt{3.72} = 1.9$ for the mock-up ignition method. See Figure 6.

This ratio is comparable to the R/r ratio for other fire test methods currently being used to regulate materials which may be involved in unwanted fires. For example, ASTM E648 (Standard Method for Critical Radiant Flux of Floor Covering Systems) has an R/r ratio of 1.1 to 1.6 [29]; ASTM E662 (Standard Test Method for Specific Optical Density of Smoke Generated by Solid Materials) has R/r ranging from 1.2 to 4.0 [30]; and ASTM E1354 (Standard Test Method for Heat and Visible Smoke Release for Materials and Products Using an Oxygen Consumption Calorimeter) has $R/r = 1.8$ for ignition delay time [31].

C. Cigarette Extinction Test Method

As previously mentioned, an ignition propensity test method need not directly simulate the upholstery material ignition process. Many flammability tests are imperfect representations of the hazard under consideration. This is because full simulation of the fire of concern is often not possible at bench scale, is too costly, or is otherwise impractical. Thus, a cigarette ignition propensity test method could measure, *e.g.*, heat release rate, were it shown to correlate with real-world ignition performance. Such a method can be useful in practice, at least over the range of cigarette designs for which it has been calibrated; and it may also be more convenient to apply.

As can be inferred from the discussion in Section II.B, the substrate requirements for a cigarette ignition propensity test method may be more readily met on a long-term basis if upholstery materials are avoided. This prompted the pursuit of alternative methods in this study. This section of the report describes the work performed in developing such a test method for the measure of cigarette ignition propensity.

1. Prior Alternative Methods

The search for a method for the evaluation of cigarette ignition propensity that is free of upholstery materials has been ongoing intermittently for over ten years. In 1981, Krasny *et al.* [32] reported a series of experiments that ultimately led to the development of a test method that employed alpha cellulose paper as a surrogate substrate. As possible indicators of cigarette ignition potential, they compared four measures of cigarette behavior to mock-up test results obtained for the same cigarettes on a variety of upholstered furniture substrates. These measures were:

- static burning rate of the cigarettes,
- surface temperature of the cigarette burn cone,
- burning behavior of the cigarettes in contact with heat sinks, and
- burning behavior of the cigarettes on alpha cellulose paper.

They concluded that weight loss rate from the cigarette/paper system was a good measure of cigarette ignition propensity, while there were shortcomings with the other three measures. Thirty commercially available cigarettes were evaluated by this test method. Reasonable agreement was found between cigarette propensity to ignite upholstered furniture substrates and weight loss rate of the cigarette/paper system. Subsequent work that was part of the TSG study [3, 3] with low ignition propensity cigarettes showed that the alpha cellulose paper would not smolder, and the cigarettes would all self-extinguish. This resulted in no recorded weight loss and, therefore, no discrimination between cigarettes.

By 1985, Norman [33] had investigated several methods for assessing cigarette ignition propensity. He used four experimental cigarettes and measured:

- the heat transfer rate to a receiver below the cigarette coal and the total heat release of a cigarette smoldering in air,

- the weight loss rate of various cigarette/substrate systems, and

- the imprint of a cigarette smoldering on a block of polyurethane foam.

While Norman could not correlate free-burn heat transfer data for cigarettes burning in air to ignition propensity and the weight loss rate was dependent on specific characteristics of the substrate, the foam imprint method appeared to hold some promise. Gann *et al.* [3] further pursued this latter method. Rather than measure the volume of the imprint, they measured the weight loss of the foam block after removal of the charred remains of the cigarette. They also recorded weight loss vs. time of the cigarette/foam system during the cigarette smoldering process. They found only a weak correlation between weight loss and cigarette ignition propensity as measured by the number of ignitions on a selected group of fabric/foam substrates.

Gann *et al.* [3] also investigated the possibility of using a heated glass plate to characterize cigarette ignition propensity. By adjusting the temperature of the glass plate, they found that cigarettes could be made to smolder their entire length. Commercially available cigarettes smoldered their entire length at ambient conditions. Low and moderate ignition propensity cigarettes would smolder their entire length only when the temperature of the glass plate was raised to between 86 and 97 °C. As with the case of the alpha cellulose paper, they noted that "No difference between the low and moderate cigarettes was evident."

2. Approaches Examined in This Study

It has been previously shown [3] that cigarettes with a wide range of ignition propensities are possible. Thus, a useful test method needs to be able to discriminate over this wide potential. Mock-up-based test methods accomplish this by using a range of substrate assemblies. The alternative methods investigated in this study accomplish this by either changing surrogate substrate characteristics or by measuring a critical cigarette property. The search for an acceptable indirect test method involved the use of various non-reactive and reactive substrates.

The Series 100 cigarettes were used in these experiments. Recall that they were effectively calibrated as to ignition propensity on both mock-ups and full-scale chairs in the TSG study. Thus they can serve here to calibrate candidate alternative test methods. The properties of these cigarettes and their ignition performance seen in the TSG study are listed in Section II.A.1.

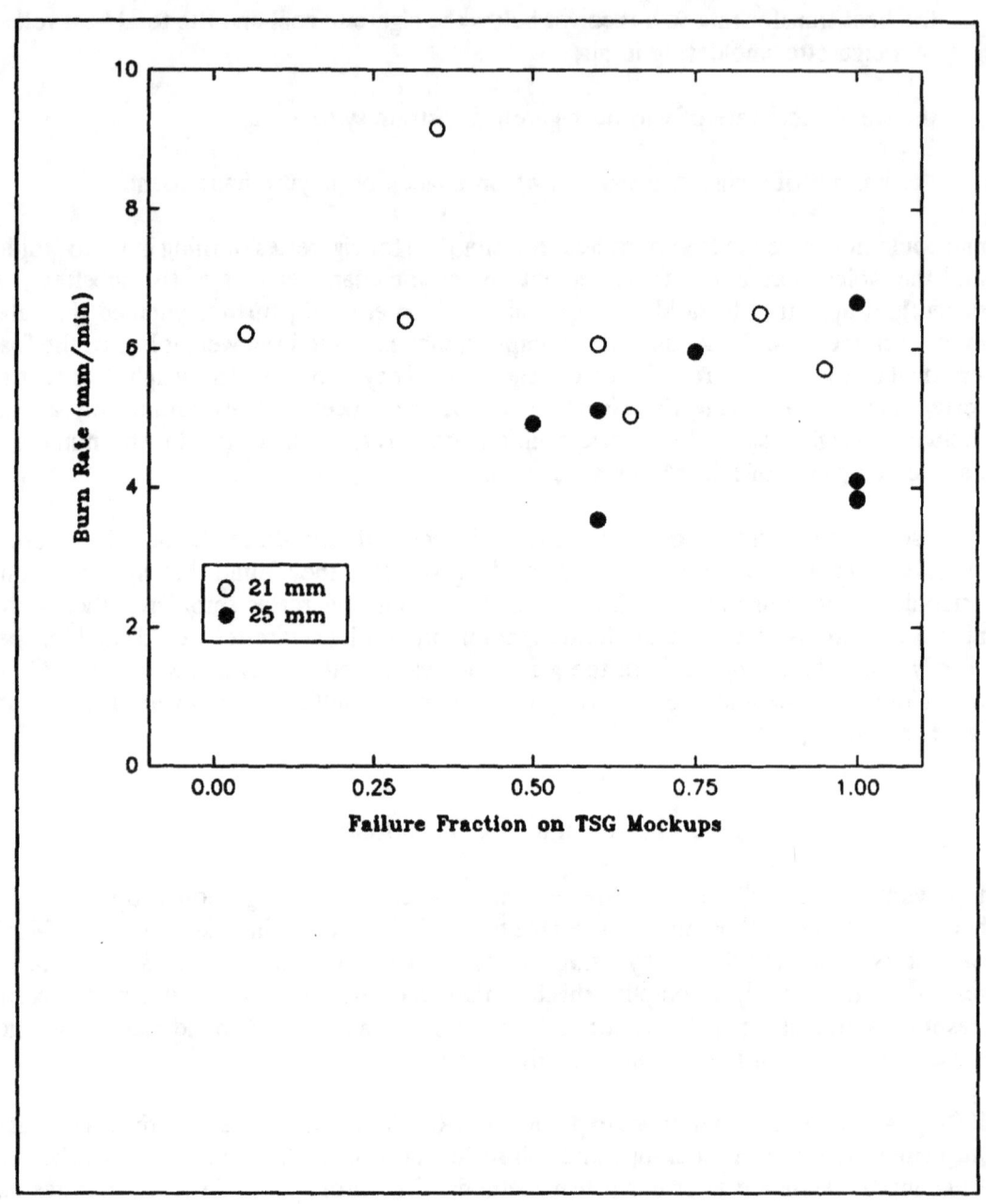

Figure 7. Free Burning Rate of Various Cigarettes Suspended in Quiescent Air as a Function of the Fraction of the TSG Mock-Up Failures.

Cigarette free-burn rate. In order to characterize baseline cigarette performance and revisit the possible use of a no-substrate test method, the burning rates of certain of the experimental cigarettes were determined under what is called "free-burn" conditions. Cigarettes were allowed to smolder while suspended horizontally in a quiescent atmosphere. All cigarettes tested in this configuration smoldered their entire length. In Figure 7, the burning rates of 16 of the TSG experimental cigarettes are plotted as a function of the ignition fractions of these cigarettes on the TSG mock-up substrates. The average burn rate for all cigarettes was 5.9 ± 1.9 mm/min. The scatter clearly exceeds the slope of a least squares line through the data. Therefore the free burn rate of a cigarette cannot be used by itself as a predictor of ignition propensity.

Non-Reactive Substrates. Non-reactive substrates are those that do not generate or absorb heat chemically when heated by a cigarette. An example of such a substrate from previous work is a glass plate. Aside from the heated glass plate previously reported, research efforts were directed at investigating three types of non-reactive substrates. These were: a bed of glass beads, a set of glass rods, and a non-woven glass fiber paper system.

Glass Beads. Figure 8 shows a schematic representation of a test setup that was used for the evaluation of glass beads as a suitable substrate for a secondary test method. The test setup consisted of a Pyrex funnel with a 125 mm diameter opening. A wire screen was suspended in the funnel such that a space of 25 mm existed between the screen and the lip of the funnel. This space was filled with glass beads, either 6 mm or 2 mm in diameter. Initial tests were conducted in a quiescent atmosphere. Subsequent tests were conducted with an imposed air flow through the glass beads, which served to disperse the flow evenly. This flow was perpendicular to the direction of cigarette smoldering. An attempt was made to find an air flow necessary to force a cigarette to smolder its entire length.

Preliminary tests were conducted on a selected subset of the TSG cigarettes at three air flows, as listed in Table 25. The air speeds through the bed of glass beads (calculated from the measured volumetric flow and the bed cross-section) were very small: 0, 0.44, 0.89 cm/min. Table 25 shows that for cigarettes 106 and 129, the number that burned their entire length increased somewhat in going from 0 cm/min to 0.44 cm/min. However, the same two cigarettes, when tested at 0.89 cm/min, showed a decrease in the number of cigarettes burning their entire length. Cigarettes 106, 129, and 130 were retested at the intermediate airflow rate of 0.44 cm/min. The retests showed somewhat erratic results, as can be seen in Table 26.

It became clear as more replicate tests were performed that cigarette performance depended a great deal on the contact characteristics of the cigarette/glass bead interface and that a reliable and reproducible contact profile could not be assured with the glass bead system. Thus this approach was not pursued further.

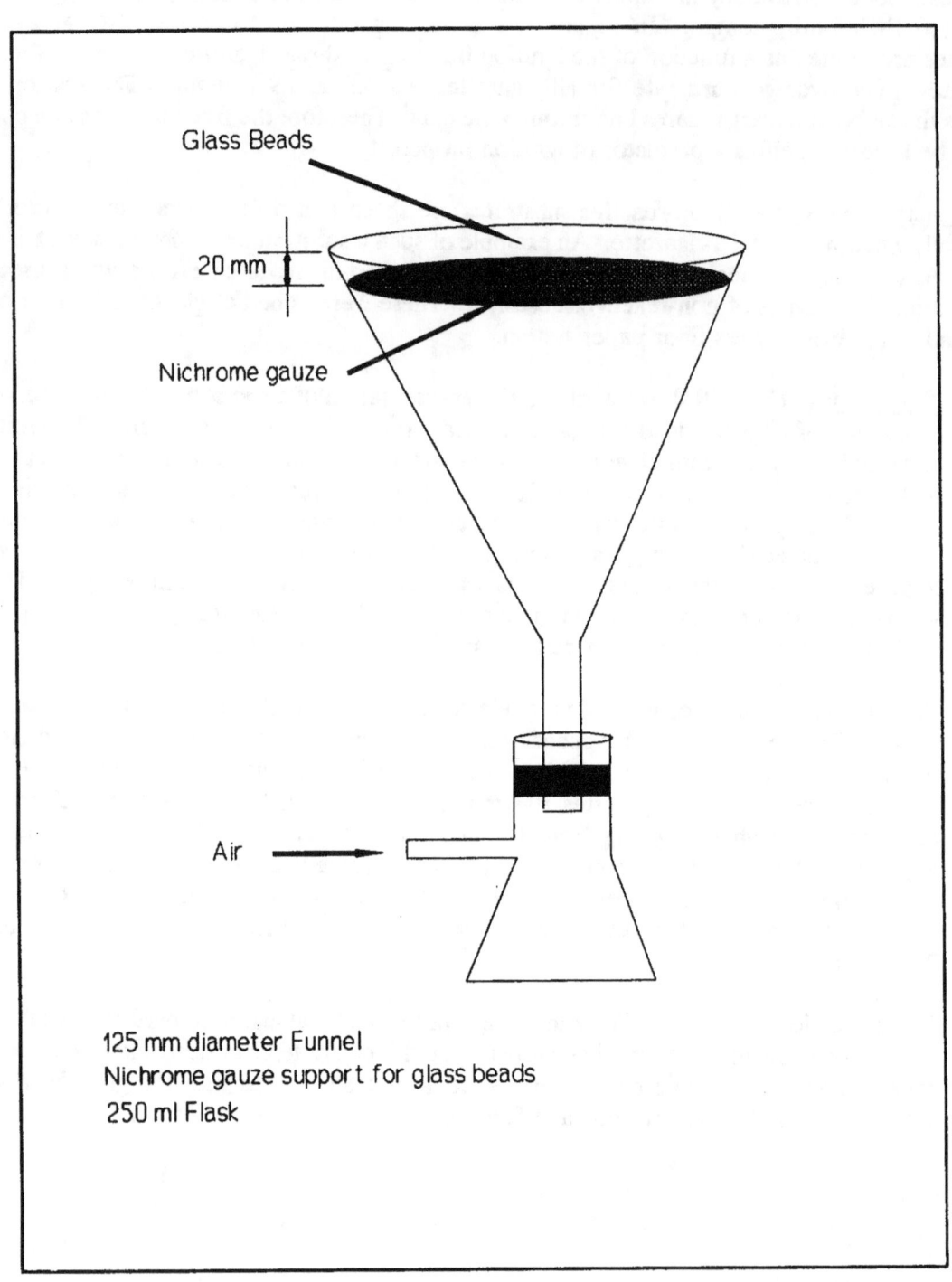

Figure 8. Schematic Representation of the Test Assembly for the Glass Bead/Rod Substrate Tests.

Table 25. Large Glass Bead Non-Reactive Substrate Test Results for Selected Cigarettes and Air Speeds

Cigarette Designation	0 cm/min		0.44 cm/min		0.89 cm/min	
	# Burn/ # Tested	Burn Time (s)	# Burn/ # Tested	Burn Time (s)	# Burn/ # Tested	Burn Time (s)
101	5/5	800 ± 60	5/5	730 ± 40		
103	5/5	590 ± 40	5/5	600 ± 20		
106	0/5	340 ± 130	5/5	665 ± 110	3/5	480 ± 200
108	5/5	570 ± 70	5/5	520 ± 95		
120	5/5	725 ± 30	5/5	690 ± 45		
129	1/5	520 ± 405	3/5	880 ± 100	1/5	390 ± 345
130	0/5	100 ± 60	0/5	635 ± 225		
131	5/5	640 ± 53	5/5	720 ± 85		
Total	26/40		33/40		4/10	

Table 26. Re-test of Selected Cigarettes on Large Glass Bead Substrate; Air Speed = 0.44 cm/min

Cigarette No.	# Burn/ # Tested	Burn Time (s)
106	1/5	540 ± 100
129	3/5	740 ± 25
130	0/5	145 ± 130

<u>Glass Rods.</u> The bed of glass beads was replaced by a pair of parallel glass rods, which were positioned on the supporting screen described above. The cigarette was placed on the rods parallel to their length, and the rods were spaced to ensure minimal contact between the cigarette and the glass rods. Air flowed upward past the smoldering cigarette as in the previous experiments. All the test cigarettes self-extinguished. This avenue of research was not pursued any further, although several test variables could have been adjusted that might have improved the discrimination capabilities of the test setup. These include: varying the temperature of the imposed airflow, replacing the glass rods with other materials, and altering the temperature of the supporting rods.

<u>Non-Woven Glass Fiber Paper.</u> In all cases, cigarettes burned their entire length when placed on a single sheet of non-woven glass fiber filter paper (with the paper suspended horizontally in air). The effective thermal inertia of this glass fiber filter paper was sufficiently low that it extracted heat from the burning cigarette less effectively than did the glass beads. While it was expected that multiple layers of this filter paper would reduce the likelihood of a cigarette burning its entire length, no such effect was observed. Instead, cigarettes such as 106 would continue burning even when supported on 10 layers of glass fiber filter paper.

This result suggested that it might be possible to use this type of filter paper to measure the heat transfer from a smoldering cigarette to a substrate material. An apparatus (Figure 9) was designed and constructed that consisted of a PMMA (polymethylmethacrylate) box with outside dimensions of 90 mm by 125 mm by 25 mm. A sheet of the glass fiber filter paper served as a cover. Three thermocouples were placed along a long diagonal of the box in an air gap approximately 12.7 mm deep between the box cover paper and the interior base of the box. The thermocouples were wired in parallel to monitor the average temperature of the three sensors.

Figure 10 shows a typical plot of the results from a single test. The graph shows the temperature-time history resulting from a complete cigarette burn. Several attempts were made to summarize and interpret the results of a single test. The best correlation to cigarette ignition propensity was an estimated heat transfer to the air gap. This involved computing a number proportional to an approximate measure of the heat content of the air in the gap, as follows:

$$Q' = T_{pk} \times \tau_{pk} \qquad (°C \cdot min)$$

where
T_{pk} = Peak temperature (°C)
τ_{pk} = Time of peak temperature (min)

Figure 11 summarizes the results for 16 different cigarettes, plotting the estimated heat content as a function of percentage mock-up ignitions from the TSG study [3]. While the data suggest that a correlation exists, the method requires further investigation and refinement. It was found that results could be dramatically affected by the specific location of the cigarette on the glass fiber filter relative to the thermocouples. Also, in order to yield repeatable results, the holder assembly needed to be cooled to ambient conditions between tests (this could take 20 to 30 minutes). Additional work along these lines was not pursued because of the success of the reactive substrate method described below.

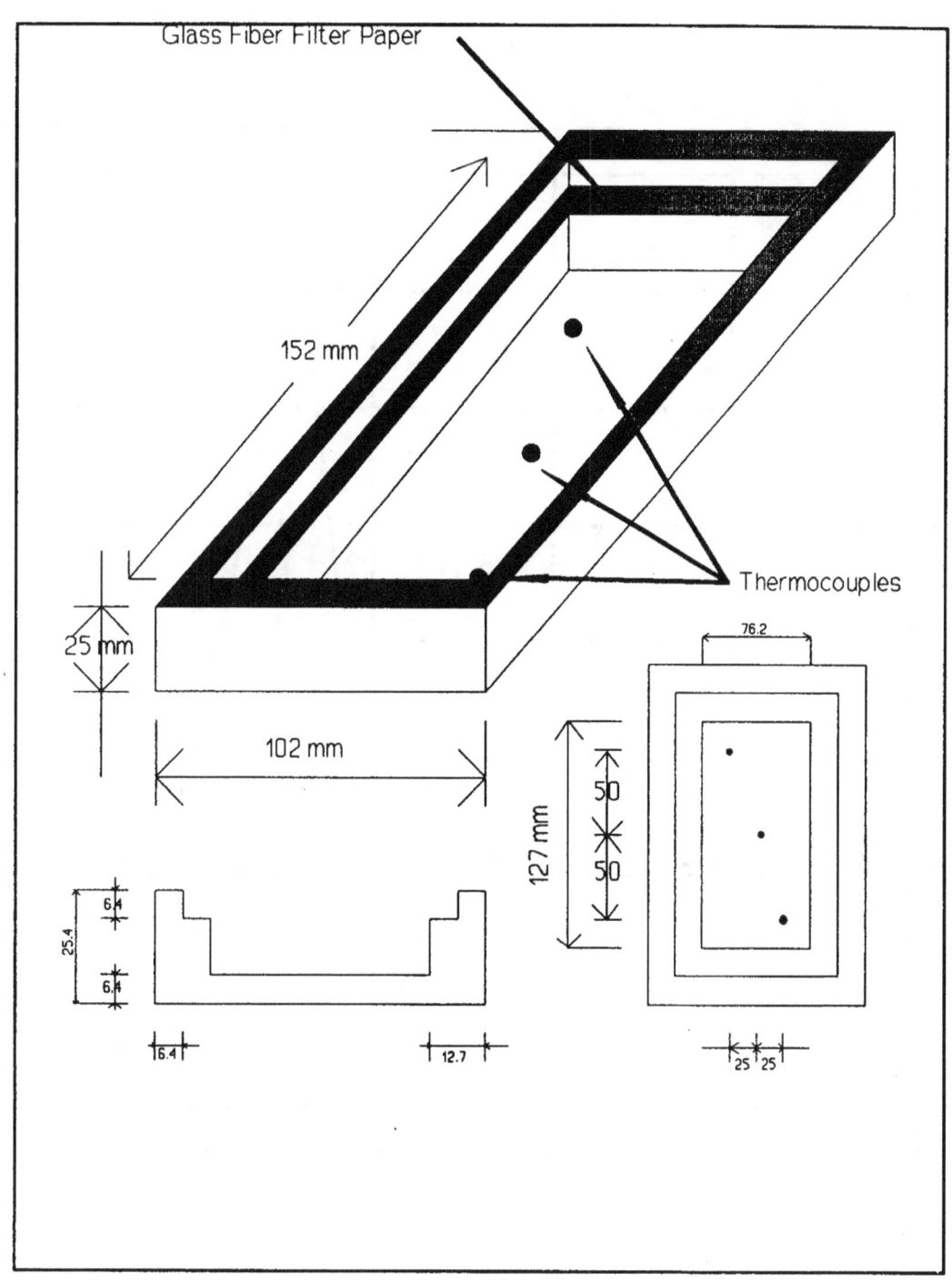

Figure 9. Drawing of the Cigarette Thermal Transfer Test Assembly.

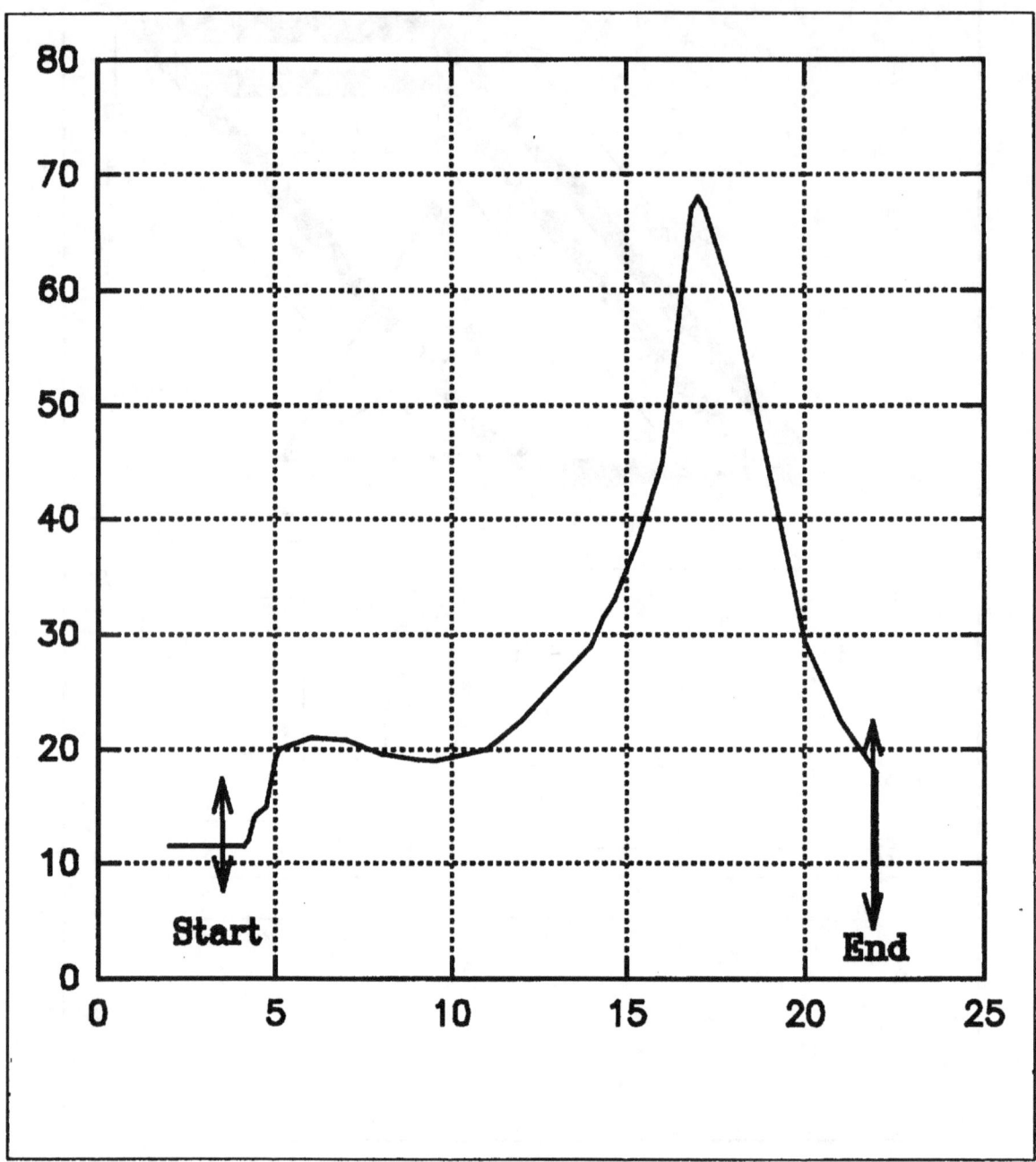

Figure 10.   Typical Average Temperature-Time Trace for a Cigarette Burning on a Glass Fiber Filter Paper in the Cigarette Thermal Transfer Test Assembly.

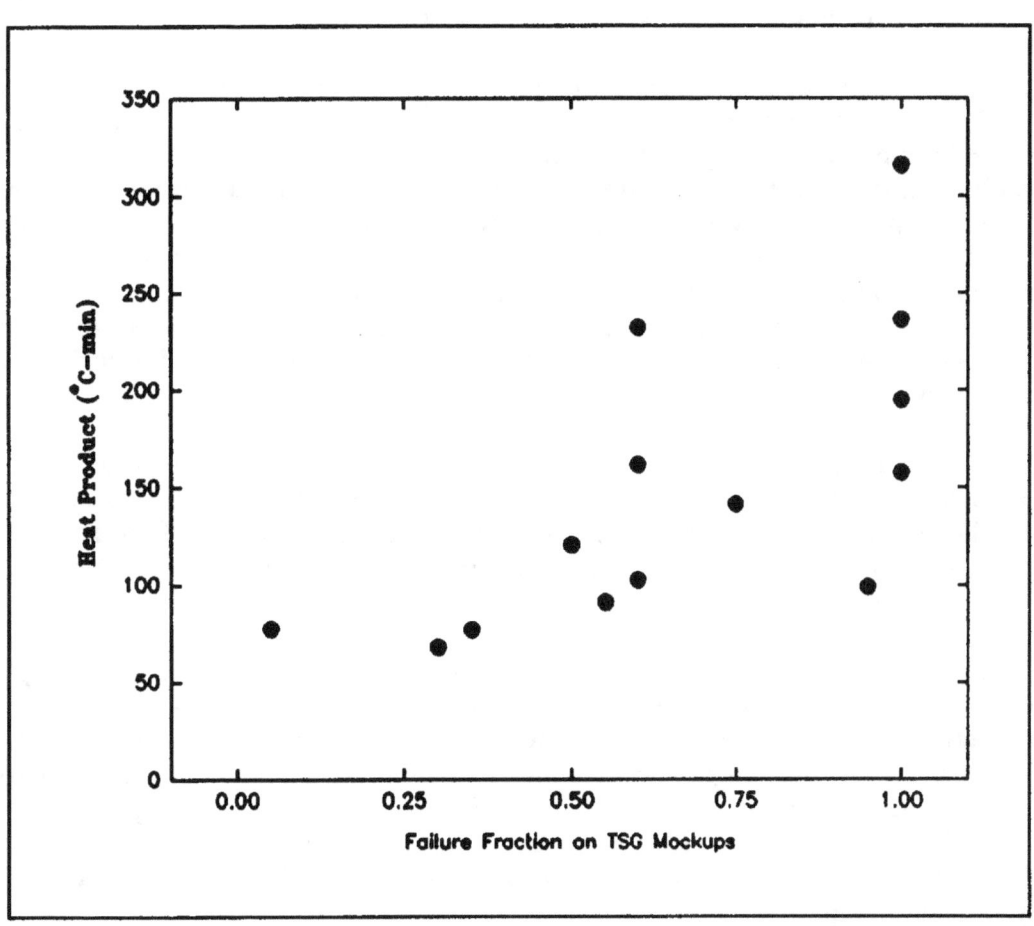

Figure 11.   Estimated Energy Transferred to a Substrate from a Smoldering Cigarette Burning in the Thermal Transfer Apparatus as a Function of the Fraction of TSG Mock-Up Failures.

Reactive Substrates. Reactive substrates undergo significant chemical change when heated by a cigarette coal. In the present study, the need was to identify reactive substrates that had advantages over the foam/fabric assemblies. Thus, these materials had to be:

- easily obtained now and in the future;
- well-characterized;
- highly uniform, both within a sample and batch-to-batch;
- smooth-surfaced; and
- available in large quantities.

After screening tests were conducted on several substrate materials (*e.g.*, lens paper, different grades of filter paper, bond paper, etc.), Whatman #2 filter paper emerged as the choice. The idea for using an alpha cellulose substrate had originated with Krasny in 1981 [32]. He used multiple layers of alpha-cellulose to support the cigarette. Since the smoldering promotor ion concentration is very small in these papers, any charring of the substrate is limited to the cigarette/substrate contact area. When the cigarette extinguishes, the substrate will not continue to smolder.

Here, it was initially assumed that the smoldering rate of a cigarette in contact with the substrate could be used as an indicator of ignition propensity. That is, the heat loss to the pyrolyzing paper would slow the burning rate of a cigarette; and at some magnitude of heat loss, the cigarette would self-extinguish. Since a single sheet of the paper is thermally-thin (small temperature gradient through its depth), more layers would extract more heat. Preliminary experiments were aimed at developing a relationship between cigarette burn rate and substrate thickness as defined by the number of filter layers in the substrate assembly. It was expected that as the number of filter paper layers increased, the burning rate would decrease. As was true for all the work performed in this program, tests were conducted in a enclosure system comparable to that used in the mock-up testing program described in the TSG report [3].

Figure 12 shows the smoldering rates of three TSG cigarettes as a function of the number of filter papers in the substrate assembly. There is a general downward trend in the data for a given cigarette. The changes, however, are not sufficient to discriminate even among cigarettes of distinctly differing ignition propensities, such as cigarette 106 with a TSG rating of 5% ignitions and cigarette 112 with a TSG rating of 100% ignitions.

However, it was noted that, as the number of filter paper layers was increased, specific cigarettes would not burn their entire length. Therefore, further work was directed at whether there was a relationship between the maximum number of filter papers in a substrate assembly that just allows a cigarette to burn its entire length and its ignition propensity as defined by the TSG ignition probabilities. A simple apparatus consisting of layers of filter paper on a metal ring was used to test the hypothesis. The preliminary data (3-5 replicates) in Figure 13 show that such a relationship does exist. The correlation covers a wide dynamic range: TSG ignition probabilities from 5% to 100% and 1 to 20 layers of filter paper.[7]

---

[7] Note that there is nothing to be gained by going to an indefinitely greater number of sheets. The heat from the cigarette can penetrate only so far in the time available. It is estimated that 25-30 sheets constitute a thermally thick medium.

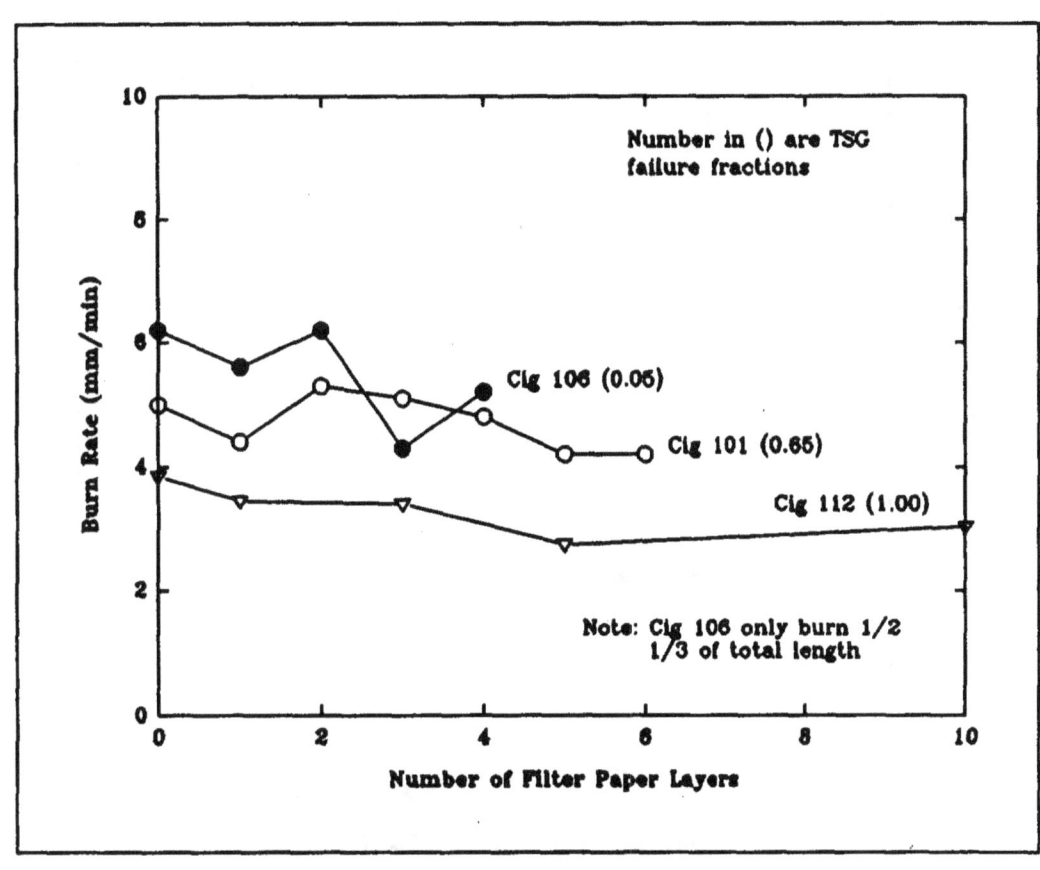

Figure 12. Smoldering Rates of Three Experimental Cigarettes as a Function of the Number of Filter Papers Making up the Substrate Assembly.

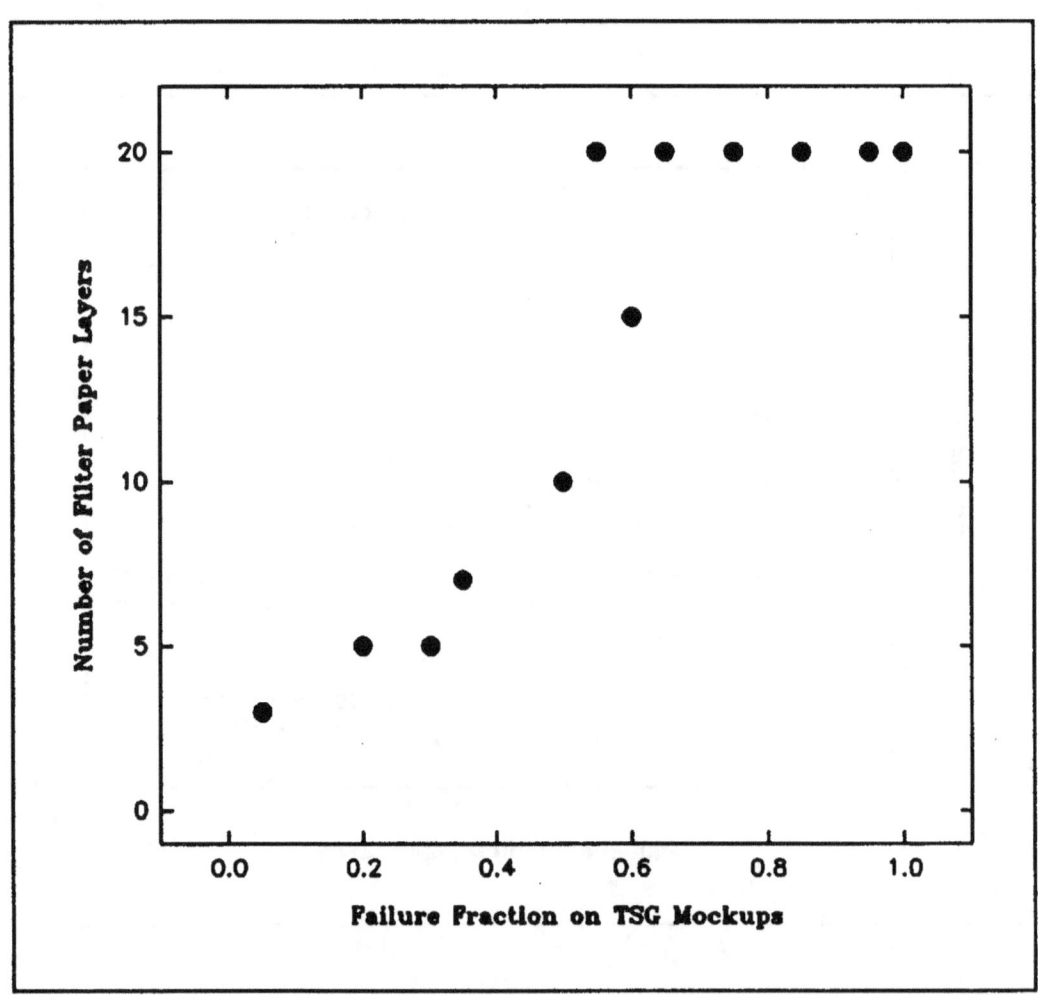

Figure 13. Number of Filter Papers Causing Extinguishment of the Cigarette as a Function of the TSG Failure Fraction.

Further testing showed that some cigarettes had a tendency to roll across the surface of the filter paper during a test. This altered the estimate of cigarette ignition propensity. A modified substrate holder assembly was developed. The holder assembly held folded filter paper such that each side of the filter paper stack was set at an angle of 20° from the horizontal. This helped ensure that cigarettes would not roll across the paper surface.

Because of the costs involved in manufacturing the 20° holder assembly and the tendency for filter paper separation to occur at the crevice joint, restraints were developed instead for the flat holder assembly. The final system is shown in Figure 14. It is composed of a brass hold-down ring with two sets of small metal rods to prevent a cigarette from rolling (yet without applying excessive pressure on the cigarette) and a plastic filter paper support structure. Each set of metal rods are spaced for a small range of cigarette diameters. As additional cigarette diameters are encountered, appropriately spaced metal rods can be added.

3. Standard Materials

Paper Substrate. The cigarette extinction test method uses multiple layers of Whatman #2 filter paper as the substrate material. It is a well-characterized material, having a well-defined porosity and filtration speed and a smooth surface finish. A single sheet has an areal density of 9.8×10^{-2} kg/m^2, with low variability (Table 27). It is made from a single material (alpha-cellulose) and should be obtainable indefinitely into the future. It is also readily available in a variety of shapes and sizes. Because of the lengths of currently manufactured cigarettes, it was felt that the standard 150 mm diameter size would be sufficient for cigarette ignition propensity measurements. The precut material reduces handling damage that might occur if each technician had to cut the paper to size.

The data in Table 27 were obtained by averaging six samples taken at random from six different boxes of Whatman #2 filter paper.

Table 27. Variability of Filter Paper Areal Density and Thickness

Box	Areal Density (kg/m^2)	Thickness (mm)
A	$(10.02 \pm .18) \times 10^{-2}$	$0.195 \pm .004$
B	$(9.74 \pm .10) \times 10^{-2}$	$0.182 \pm .003$
C	$(9.80 \pm .13) \times 10^{-2}$	$0.181 \pm .005$
D	$(9.58 \pm .13) \times 10^{-2}$	$0.184 \pm .005$
E	$(9.71 \pm .18) \times 10^{-2}$	$0.183 \pm .004$
F	$(9.68 \pm .32) \times 10^{-2}$	$0.187 \pm .005$
Overall:	$(9.76 \pm .15) \times 10^{-2}$	$0.185 \pm .005$

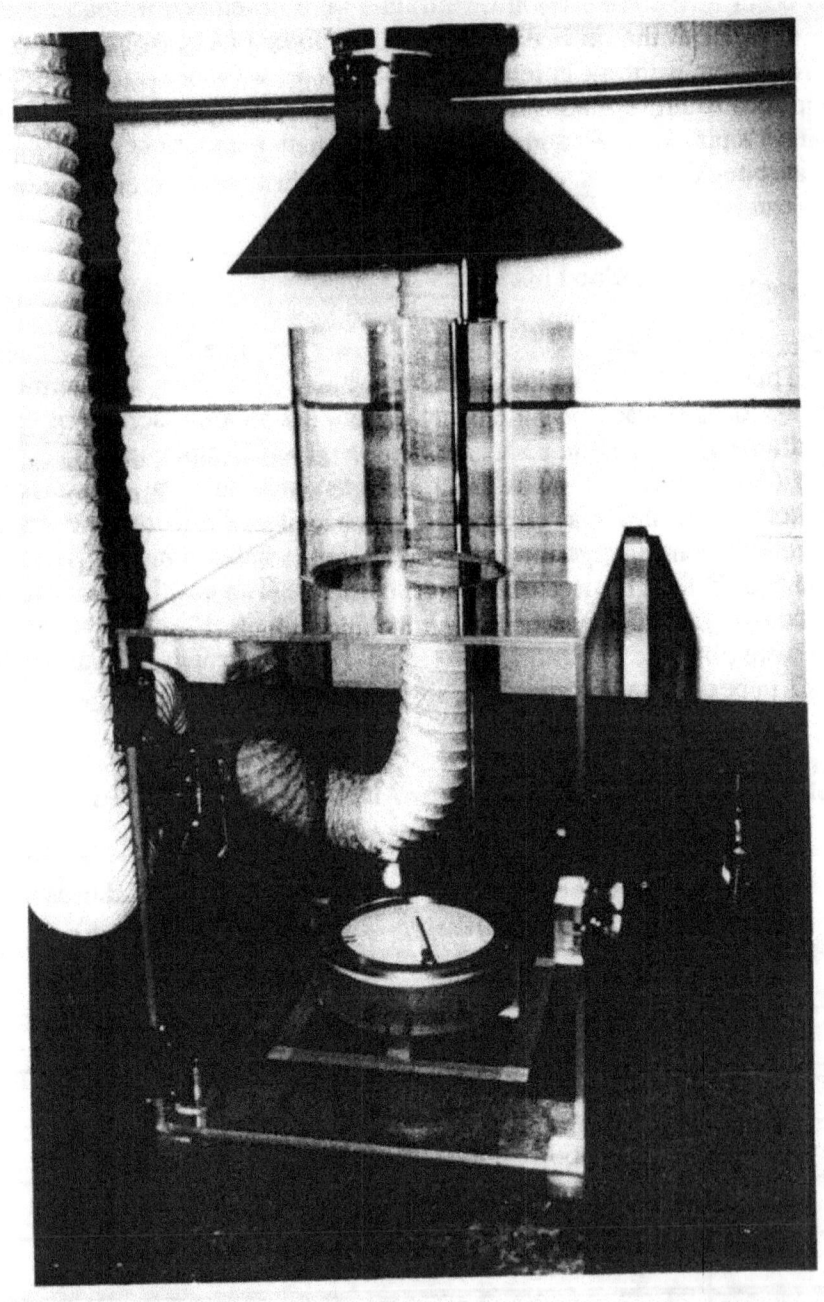

Figure 14. Photograph of a Test Chamber Containing a Mock-Up Assembly and a Cigarette.

Substrate Description. The test method concept originally involved determination of the actual number of filter paper layers necessary to just allow the cigarette to burn its complete length. To reduce the testing burden on the participating laboratories as well as to reduce the amount of filter paper used in each cigarette evaluation, the interlaboratory evaluation was performed with three specific numbers of layers. This also enabled using a statistical design comparable to the one used for the mock-up ignition test method. The substrates comprised 3, 10, and 15 layers of Whatman #2 filter paper. In practice, the original concept may have application as well.

4. Enclosure Design

This test method simply replaces the fabric/foam substrate with an alternative substrate assembly. Since it has been demonstrated that the enclosure used in the mock-up ignition test method adequately protects the cigarette-substrate system from laboratory induced air flows, that enclosure was also employed in the cigarette extinction test method.

5. General Description of the Test Method

Appendix E gives a detailed description of the cigarette extinction test method. In brief, the test method measures whether a type of cigarette continues smoldering after being placed on substrate assemblies that have different thermal absorptivities. The appropriate number of layers of Whatman #2 filter paper are mounted on the support structure described above and placed in the enclosure. The cigarettes and substrate assemblies are conditioned at a relative humidity of 55 ± 5% and a temperature of 23 ± 3 °C. Cigarettes are ignited and pre-burned to a 15 mm mark as described for the Mock-Up Ignition Test Method. The principal determination is whether the cigarette burns its full length or not.

6. Interlaboratory Study of the Test Method

a. Participants and Procedures

The nine laboratories participating in this phase of the interlaboratory study were the same as previously listed for the main ILS. See the list of participants in Section II.B.8.e. The general test protocol for this phase of the ignition propensity study followed that outlined for the ILS of the mock-up extinction method. The only major differences were that (a) a different method was being studied and (b) fewer replicates were performed. The latter was proposed since the substrate variability, thought to be a potential factor in the precision in the mock-up method, was minimal here. The reduced test plan used only that portion of the plan specified for each laboratory in the first week of testing during the main ILS. The following outlines key parameters pertaining to each laboratory in this study:

- 5 cigarettes
- 3 filter paper substrates (3, 10 and 15 sheets thick)
- 16 replicates per cigarette per substrate
- 2 operators
- 4 test chambers

- 1 week test period

Each laboratory received 25 boxes of filter paper, four plastic filter paper supports and brass hold-down rings, 100 coded cigarettes, plus instructional manuals and lab workbooks for each operator. Each laboratory already had in their possession test chamber enclosures, cigarette lighters, cigarette holders, etc. from the mock-up ignition interlaboratory test program. Similar procedures were followed to ensure timely arrival of test data to NIST via the FAXing of daily summary sheets. At the end of the one-week test program, workbooks and data disks were returned to NIST. These were reviewed for consistency between the workbooks, disk data files, and daily summary sheets. Discrepancies were noted and resolved to ensure an accurate set of data files from each laboratory.

b. Analysis of Results

Raw Data. As was done for the previous interlaboratory studies, the data for the cigarette extinction test method were organized into a single computer file for analysis. For this study, the resulting computer file contained 2160 lines of data, corresponding to 240 ignition results per laboratory for each of 9 laboratories. The 240 results per laboratory arise from 16 tests of each of 5 cigarettes on each of 3 substrates.

As was the case for the main ILS of the Mock-Up Ignition Method, there were a few cases in the ILS of the Cigarette Extinction Method where laboratories ran some test replications on the wrong mock-up configurations. This resulted in some cases where the number of results reported for a given cigarette type and mock-up configuration differ, again by ± 2, from the desired number of 16 replications.

The raw data on each line of the computer file represent essentially the same set of thirteen variables described in Table 20, with the familiar changes that (a) the LAB variable ranges from 1 to 9, and (b) the OPERATOR variable codes the two operators as number "1" and number "2." These differences were discussed in more detail in Section II.B.8.c. Another difference from Table 20 is that the TST_BLK variable was not used because this study was done in a single week of testing.

A summary of the test results, by LAB, CIG_TYPE and SUBSTRAT, is presented in Table 28. The identifying numbers for the laboratories are the same as those used in Table 22. The test results for the extinction method were reported in two categories, full-length burns and self-extinguishments.

Table 28. Summary of Test Results for Interlaboratory Study of Cigarette Extinction Method

Cigarette Type	Substrate (No. of Layers)	Laboratory	Test Results	
			Full-Length Burns	Self-Extinguishments
501	3	1	16	0
		2	16	0
		3	16	0
		4	16	0
		5	16	0
		6	16	0
		7	16	0
		8	16	0
		9	16	0
	10	1	16	0
		2	16	0
		3	16	0
		4	16	0
		5	16	0
		6	16	0
		7	16	0
		8	16	0
		9	16	0
	15	1	16	0
		2	16	0
		3	16	0
		4	16	0
		5	16	0
		6	16	0
		7	16	0
		8	16	0
		9	16	0

Cigarette Type	Substrate (No. of Layers)	Laboratory	Test Results	
			Full-Length Burns	Self-Extinguishments
503	3	1	16	0
		2	16	0
		3	16	0
		4	16	0
		5	16	0
		6	16	0
		7	16	0
		8	16	0
		9	16	0
	10	1	16	0
		2	16	0
		3	16	0
		4	18	0
		5	16	0
		6	16	0
		7	16	0
		8	16	0
		9	16	0
	15	1	16	0
		2	16	0
		3	16	0
		4	14	0
		5	16	0
		6	16	0
		7	16	0
		8	16	0
		9	16	0

Cigarette Type	Substrate (No. of Layers)	Laboratory	Test Results	
			Full-Length Burns	Self-Extinguishments
529	3	1	7	9
		2	8	8
		3	11	5
		4	9	7
		5	8	8
		6	10	6
		7	8	8
		8	8	8
		9	13	3
	10	1	0	16
		2	1	15
		3	0	16
		4	0	16
		5	0	16
		6	1	15
		7	0	16
		8	1	15
		9	5	11
	15	1	0	16
		2	0	16
		3	0	16
		4	0	16
		5	0	16
		6	0	16
		7	0	16
		8	1	15
		9	2	14

Cigarette Type	Substrate (No. of Layers)	Laboratory	Test Results	
			Full-Length Burns	Self-Extinguishments
530	3	1	2	14
		2	0	16
		3	0	16
		4	0	16
		5	0	16
		6	2	14
		7	1	15
		8	1	15
		9	2	14
	10	1	0	16
		2	0	16
		3	0	16
		4	0	16
		5	0	16
		6	0	16
		7	0	16
		8	0	16
		9	0	16
	15	1	0	16
		2	0	16
		3	0	16
		4	0	16
		5	0	16
		6	0	16
		7	0	16
		8	0	16
		9	0	16

Cigarette Type	Substrate (No. of Layers)	Laboratory	Test Results	
			Full-Length Burns	Self-Extinguishments
531	3	1	16	0
		2	16	0
		3	16	0
		4	16	0
		5	16	0
		6	16	0
		7	15	1
		8	16	0
		9	16	0
	10	1	14	2
		2	16	0
		3	15	1
		4	13	3
		5	16	0
		6	16	0
		7	15	1
		8	16	0
		9	15	1
	15	1	10	6
		2	13	3
		3	14	2
		4	15	1
		5	15	1
		6	14	2
		7	16	0
		8	15	1
		9	15	1

A graphical display of the data in Table 28 is shown in Figure 15, where, for each cigarette (by columns) and substrate (by rows) the proportion of full-length burns is represented by a vertical bar for each laboratory. The cigarette types are shown from left to right in order of decreasing ignition propensity, as determined by the results of the main interlaboratory study of the Mock-Up Ignition Test method (see Figure 4). A comparison of Figure 15 with Figure 4 shows that, except for cigarettes 501 and 503 which are tied in Figure 15, the relative positioning of the cigarettes was the same in both the studies. The three mock-up configurations are shown as rows in the figure, with the greatest heat-sink substrate (15 layers of filter paper) as the top row and the least heat-sink substrate (3 layers of filter paper) as the bottom row of the figure. The stronger the heat sink ability of the substrate, the more difficult it is for a cigarette to burn its full length on that substrate.

Auxiliary Variables. A few statistical procedures were run to check for any interesting or large effects on the test results associated with the available auxiliary variables. No statistically significant effects were found associated with the temperature and humidity variables over the ranges occurring in these tests. For comparison with the previous interlaboratory studies, box plots showing the ranges of the temperatures and humidities during testing are given in Figure 16.

The Cochran-Mantel-Haenszel test was used to check for significant effects on ignition results due to the discrete variables. No significant effect was found due to OPERATOR; however, the tests on AMPM and CHAMBER did achieve statistical significance. For the AMPM variable, lab 7 showed a significant effect. Detailed study of the data for lab 7 revealed that the significant difference that was picked-up by the overall Cochran-Mantel-Haenszel test procedure was entirely due to the results for cigarette 529 on the 3-layer substrate. In that case, there were 7 ignitions and 1 self-extinguishment in the AM and just the opposite, 1 ignition and 7 self-extinguishments, in the PM. Using Fisher's Exact Test for the resulting 2×2 contingency table yields a significance probability of 0.01. No other aspect of the data for this case looks unusual, so the decision was made to accept the data as-is. It is relevant to note that, since a large number of significance tests were conducted on this data set, one would expect a few cases to show statistical significance simply due to the expected amount of random variation in the data.

For CHAMBER, the statistically significant effect flagged by the Cochran-Mantel-Haenszel test was for lab 3 only. Within the data for lab 3, the significance was caused by the results for the 16 tests of cigarette 529 on the 15-layer substrate. The significance probability for this case was 0.01 by Fisher's Exact Test. Again, nothing else unusual was found regarding the data in question, and the existing data were used in the repeatability and reproducibility summary without modification.

Primary Variables. Statistical tests were carried out to examine whether these interlaboratory study data reveal differences between the labs, cigarette types and substrates. For labs, there is a statistically significant difference only for cigarette 529 on the 10-layer substrate. The fact that only one case showed a difference for the Cigarette Extinction Method is at least partly due to the fact that only 16 replications were done per laboratory, rather than the 48 in the main ILS of the Mock-Up Ignition Method. With fewer data, fewer significant differences are likely to be found, even if the long-run differences are about the same.

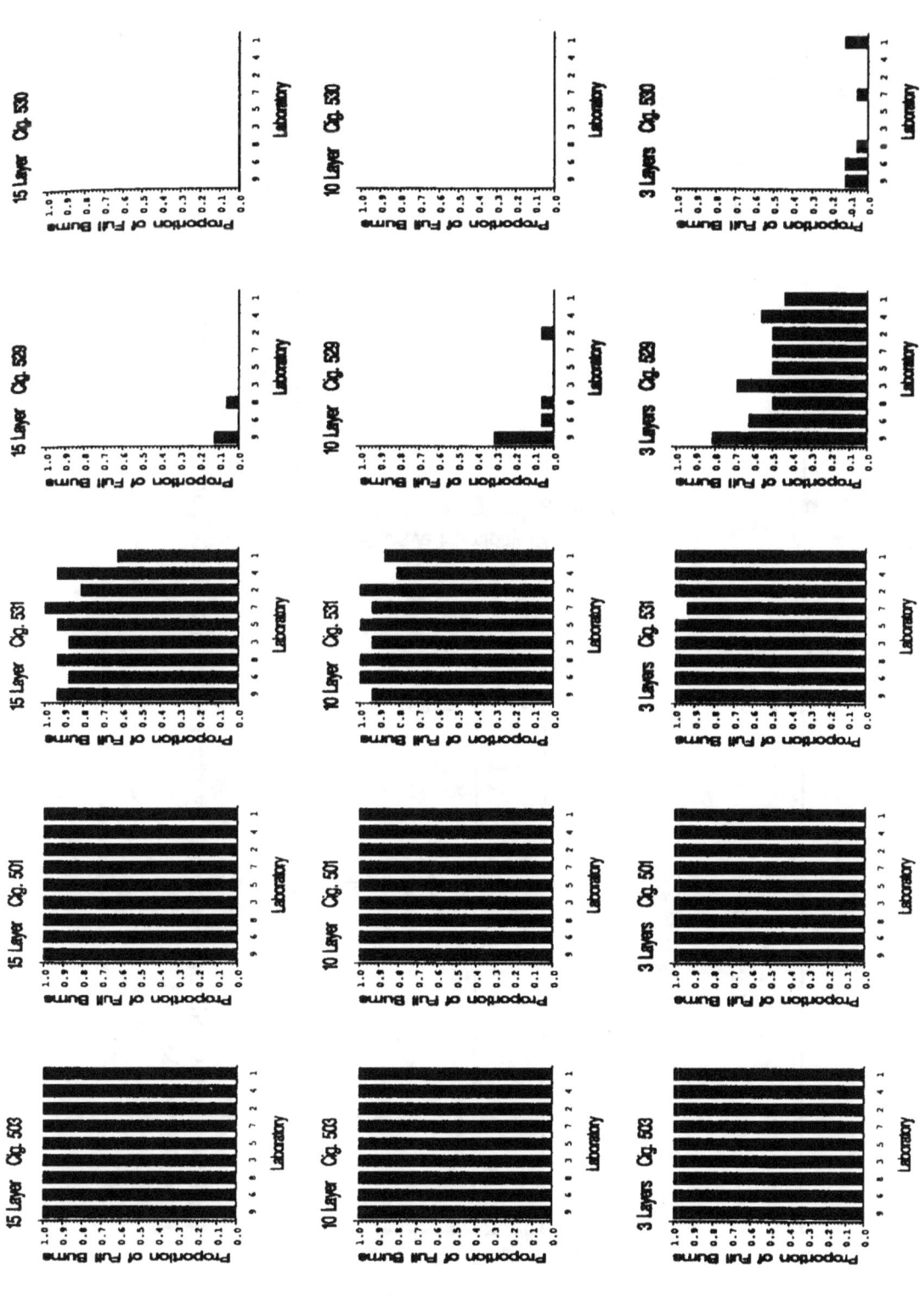

Figure 15. Comparison of Full Burn Rates for the Interlaboratory Study of the Cigarette Extinction Test Method. The 15 plots in the figure correspond to the 5 cigarette types tested, by columns, and the 3 test substrates (15, 10, and 3 filter layers), by rows. In each component plot, the vertical bars represent the proportions of full burns obtained by each of the 9 participating laboratories.

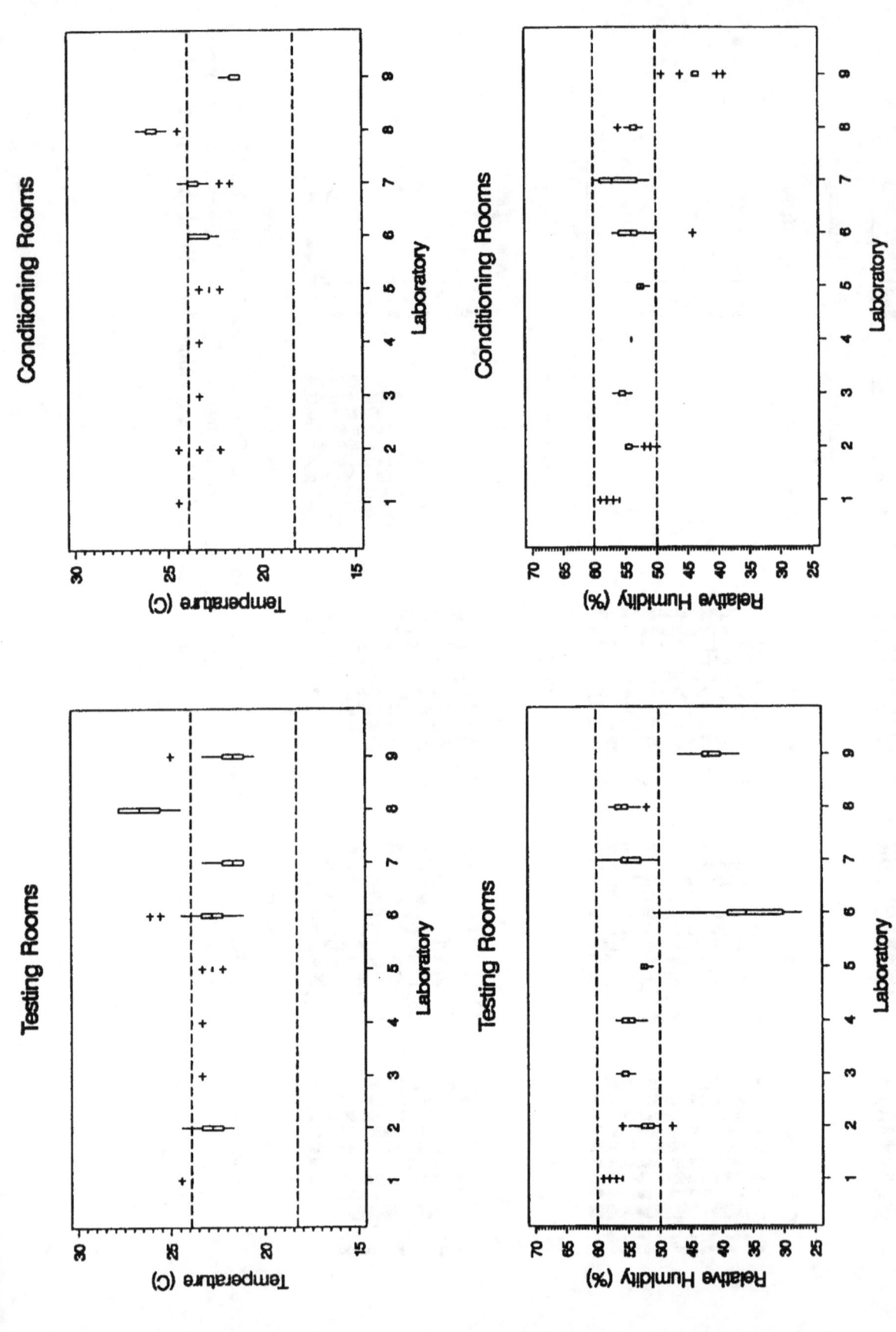

Figure 16. Environmental Conditions Reported by the Laboratories Participating in the Interlaboratory Study of the Cigarette Extinction Test Method.

Except for the fact that cigarettes 503 and 501 gave identical results (100% full-length burns for all labs and all substrates) the Cochran-Mantel-Haenszel procedure showed that the cigarettes differ from each other: 501 and 503 having higher full-length burn proportions than 531, which is higher than 529, which is higher than 530. Similarly, the observed differences in full-length burn proportions between the three substrates are all statistically significant according to the Cochran-Mantel-Haenszel procedure.

Repeatability and Reproducibility. The summary shown in Tables 29 and 30 follows the same methodology described previously for Tables 23 and 24. The only difference in detail is that, in the interlaboratory study for the Cigarette Extinction Method, the number of replications per lab was $m=16$, compared to $m=48$ in the Mock-Up Ignition Method. Therefore, in Table 29, the repeatability standard deviation is calculated as $S_r = [\bar{p}(1-\bar{p})/16]^{1/2}$.

The model for extra-binomial variation used previously for the Mock-Up Ignition Method ILS was also applied to these data. The observed relation between the reproducibility variance, S_R^2, and the repeatability variance, S_r^2, for the Cigarette Extinction Method ILS is shown graphically in Figure 17. The slope of the least squares line in the Figure is 1.146. Setting this value equal to the heterogeneity factor, $[1 + \varphi(16-1)]$, and solving for φ, yields the estimate $\varphi=0.0097$. The resulting summary of repeatability and reproducibility limits is shown in Table 30.

Because the estimate of the correlation parameter, φ, was somewhat smaller for the Cigarette Extinction Method, where $\varphi=0.0097$, compared to the Mock-Up Ignition Method, where $\varphi=0.058$, the values of the reproducibility limits (R) are somewhat smaller for the Cigarette Extinction Method, Table 30, compared to the Mock-Up Ignition Method, Table 24. In contrast, the repeatability limits (r) are exactly the same in Tables 24 and 30 because both use the same formula for S_r, as given by Equation (1) of Section II.B.8.e.

Table 29. Observed Repeatability and Reproducibility Standard Deviations for Cigarette Extinction Method Interlaboratory Study
$m=16$ Replications per Laboratory

Cigarette I.D.	Substrate (No. of Layers)	Average Proportion of Full Length Burns	Repeatability S.D. S_r	Reproducibility S.D. S_R
501	3	1.000	0	0
501	10	1.000	0	0
501	15	1.000	0	0
503	3	1.000	0	0
503	10	1.000	0	0
503	15	1.000	0	0
529	3	0.569	0.124	0.119
529	10	0.056	0.057	0.101
529	15	0.021	0.036	0.044
530	3	0.056	0.057	0.058
530	10	0.000	0	0
530	15	0.000	0	0
531	3	0.993	0.021	0.021
531	10	0.944	0.057	0.066
531	15	0.882	0.081	0.110

Table 30. Cigarette Extinction Method: Calculated Repeatability (r) and Reproducibility (R) Limits for Various Assumed Numbers of Replications (m) and Full-Length Burn Proportions (p)

p	m = 16		m = 32		m = 48		m = 96		m = 9600	
	r	R	r	R	r	R	r	R	r	R
0.05 or 0.95	0.15	0.16	0.11	0.12	0.09	0.11	0.06	0.09	0.006	0.06
0.10 or 0.90	0.21	0.22	0.15	0.17	0.12	0.15	0.09	0.12	0.009	0.08
0.20 or 0.80	0.28	0.30	0.20	0.23	0.16	0.20	0.11	0.16	0.011	0.11
0.30 or 0.70	0.32	0.34	0.23	0.26	0.19	0.22	0.13	0.18	0.013	0.13
0.40 or 0.60	0.34	0.37	0.24	0.28	0.20	0.24	0.14	0.19	0.014	0.14
0.50	0.35	0.37	0.25	0.28	0.20	0.24	0.14	0.20	0.014	0.14

m: Assumed number of replications per laboratory
p: Assumed long-run proportion of full-length burns for cigarette and substrate combination under test
r: Repeatability limit = $2.8S_r$, where S_r is calculated as in Table 24.
R: Reproducibility limit = $2.8S_R$, where S_R is calculated as in Table 24, with $\varphi = 0.0097$

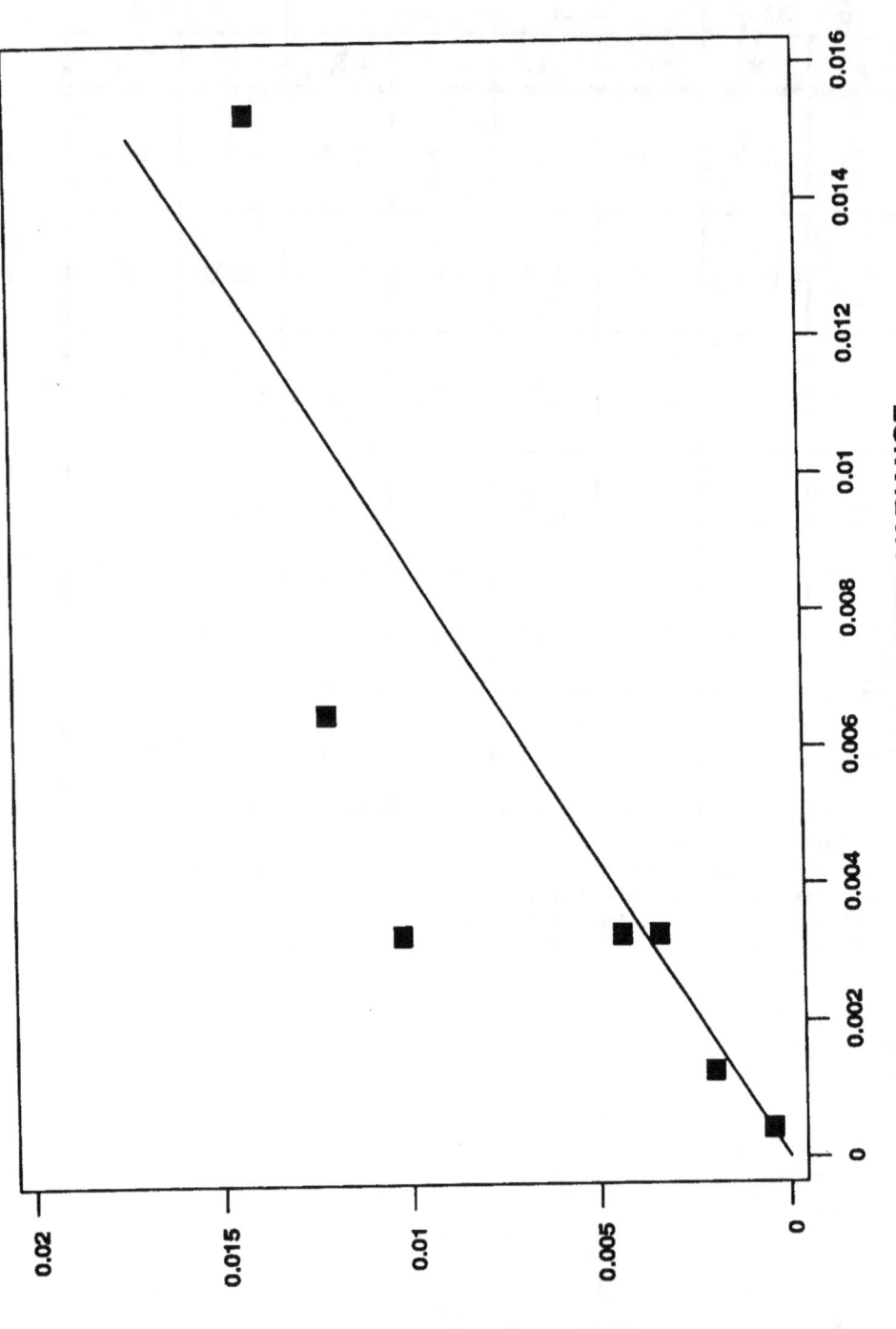

Figure 17. Empirical Relation of Reproducibility Variance, S_R^2, to Repeatability Variance, S_r^2, Based on Data in Table 29 from the Interlaboratory Study of the Cigarette Extinction Test Method. The equation of the least squares line shown is $S_R^2 = (1.146)S_r^2$.

III. CONSIDERATIONS REGARDING THE USE OF THE TWO TEST METHODS

A. Mock-Up Ignition Test Method

The mock-up method developed here broadly meets the criteria described in Section I for an acceptable test method. It is a performance-based method that employs a cigarette/substrate combination bearing a strong (although not perfect) similarity to the real-world fire safety hazard. The relation between the test results and real upholstered chair ignition behavior is traceable through the use of cigarettes calibrated in the TSG study. The test output is quantitative and provides differentiation among cigarettes of varied ignition propensity. Through choice and control of materials it should provide a stable standard of performance for the foreseeable future.

As is generally the case with fire tests, this method has potential limitations that are a consequence of incomplete knowledge of the real-world scenarios. First, in the apparatus, the ambient atmosphere is perturbed only by the cigarette plume. This case is believed to be a highly relevant analog for real-world accidental ignitions occurring in a chair crevice. As noted previously, if further information on real-world ignitions indicates a significant fraction occurring in external air flow conditions *at the ignition location*, it may be appropriate to supplement the results of the current method with those obtained in the presence of a comparable flow. This would require a method development process comparable to that described in this report. A second limitation is the small number of upholstery substrates used to relate mock-up behavior to real-world chairs [3]. It is presumed that this correlation is representative of the aggregate furniture market. The existence of this correlation virtually assures that there will be some real-world benefit in moving toward cigarettes which perform well in this test method. Should sufficient evidence emerge in the future that a large fraction of the furniture at risk does not follow the correlation that was demonstrated in the TSG study, it may be appropriate to replace one or more of the Mock-Up Ignition Test Method substrates.

The interlaboratory study demonstrated the level of lab-to-lab reproducibility one can expect of this method. Table 24 above shows that this level cannot be made substantially greater with a very large number of replicates; conversely, this level does get significantly less desirable if the number of replicates is reduced substantially below 48. The achievable lab-to-lab reproducibility is an appropriate measure of how finely a test method can differentiate among test subjects for regulatory purposes. It is apparent, then, that the mock-up method cannot make fine distinctions in ignition propensity among cigarette designs. With 48 replicates on a given mock-up, the proportion of ignitions obtained by two separate laboratories can be expected to differ by up to about 0.4. *This places a limit on the degree of resolution possible for regulatory use of the method.* Finer distinctions than this could be made only within a single laboratory, presumably for product development purposes. In that case, a number of replicates greater than 48 would appreciably improve the differentiation; see Table 24.

Three mock-ups were included in the interlaboratory study and all were found to differ significantly in ease of ignition. The results in Figure 4 above show the response of this set of mock-ups to a broad spectrum of experimental cigarette designs, though all are of a conventional construction. Given the limitation above on the resolution of the test in differentiating ignition propensity, it is apparent that cigarettes of high ignition propensity gave effectively the same response on two out of three of the mock-ups. A cigarette of low ignition propensity also gave effectively the same response on two (different) mock-ups. This does not mean that one mock-up can be omitted from the entire test set, because the duplication of response occurs on differing mock-up pairs with differing cigarettes. It does, however, suggest the possibility of the need for fewer replicates on at least one mock-up, and possibly on two, for some cigarettes.

Figure 4 shows that the duck #4 mock-up is, for the set of cigarettes examined, consistently harder to ignite than the other two mock-ups; and, in turn, the duck #6 mock-up is harder to ignite than the duck #10 mock-up. If a cigarette were tested first on the duck #4 mock-up and gave all ignitions in 48 replicates, it should be possible to do fewer replicates on the other two mock-ups to verify that there were no unexpected reversals in ignition behavior. This could provide significant labor savings on what is otherwise a rather labor-intensive test protocol.

Performing fewer than 48 tests will result in some loss of information, and a corresponding increase in statistical uncertainty regarding the long-run ignition rate that would be observed for those mock-ups. The (within-lab) statistical uncertainty, for any number of replications, can be quantified by use of confidence bounds on the ignition probabilities for a given cigarette on a given mock-up, as follows.

If a cigarette is tested on, *e.g.*, the duck #4 mock-up, and all 48 replications result in ignitions, then a 95% lower confidence bound on the long-run ignition probability is $(1-0.95)^{1/48} = 0.94$. Speaking loosely, one can be 95% confident that the ignition probability (in the same laboratory) is greater than about 0.94. With that result, it may be sufficient for some purposes to know that the ignition probability on the duck #6 and duck #10 mock-ups would be about 0.6 or higher, since the degree of interlaboratory resolution is about 0.4, as noted above. If so, then only $n=6$ runs would be required, because $(1-0.95)^{1/6} = 0.61$. That is, if 6 runs were conducted resulting in 6 ignitions, then the 95% lower confidence bound for the ignition probability would be 0.61. There can be, of course, no guarantee that 6 runs on a more ignitable mock-up would *necessarily* result in 6 ignitions. If one or more non-ignitions did occur, it would be appropriate to run a full set of 48 replications for each of the three mock-ups.

Table 31 shows the relationship between the number of runs and the corresponding lower confidence bounds in the case of 100% ignitions. The table is useful for comparison and to help decide whether the increased uncertainty due to running fewer than 48 replications is acceptable for a particular purpose.

These same arguments can be made for any cigarette and mock-up combination where a 100% response, either ignitions or non-ignitions, is obtained. It is the choice of the regulator whether this trade-off is implemented in any adopted test method.

Table 31. 95% Lower Confidence Bounds for the Long-Run Ignition Probability Assuming that n Tests Result in n Ignitions

n = Number of Runs	Lower Confidence Bound
4	0.47
8	0.69
12	0.78
24	0.88
36	0.92
48	0.94

Informal reports of in-progress cigarette industry studies imply that some upholstery fabrics will respond to contact with a lit cigarette in a substantially different manner from that seen with the cotton ducks used in this method. Even if this is verified, the possible results of employing the Mock-Up Ignition Test Method developed here are as follows:

- Some cigarette designs will produce fewer ignitions (than the current market cigarettes) both in the test and when in contact with furniture containing fabrics which behave like the cotton ducks used here. The test method in this case is a true indicator of less fire-prone cigarettes.

- Some cigarette designs will produce fewer ignitions in the test, but will not produce a reduced number of ignitions when in contact with furniture containing fabrics which is dissimilar in response to cotton ducks. It seems implausible that such designs would show greater real-world ignition propensity than do current commercial cigarettes.

- Still other cigarette designs will produce a number of ignitions, both in the test method and when in contact with furniture containing fabrics which behave like the cotton ducks used here, comparable to current commercial cigarettes. For these, the test method is again a true indicator of expected fire performance.

- Some designs will produce a number of ignitions in this test that are comparable to current commercial cigarettes, but will produce a reduced number of ignitions when in contact with furniture containing fabrics which is dissimilar in response to cotton ducks. These designs would not likely be pursued; but, if they were, they would unobtrusively reduce fire losses.

The result of the second and fourth occurrences is an uncertainty in the *degree* to which real-world fire losses are reduced.

At present there are insufficient data available to estimate what fraction of real-world furniture might contain fabrics differing substantially (*i.e.*, beyond the reproducibility of the test method) from cotton ducks in their ignition behavior. If further data become available indicating that such fabrics are a significant fraction of the real-world population, it would be an option to supplement the results of cigarette testing using this method with results based on other carefully-chosen fabrics.

B. Cigarette Extinction Test Method

An analog to most of the discussion in the preceding section applies to this method as well. The potential limitation on the imposition of an external air flow manifests itself in both methods. Since the filter paper is a surrogate material, the pertinent consideration is the degree of differentiation of cigarette ignition propensity. This is reflected in the numbers of layers selected for the test substrates. For instance, were there an interest in better discrimination among cigarettes of high ignition propensity than is shown in Figure 15, one might be inclined to select 20 or 25 layers to replace the 15 in the first substrate. However, limited data indicate that this increase has no effect on the burning behavior of cigarettes in this test series. Thus, this method is less appropriate than the mock-up method for distinguishing initial progress from current market cigarettes toward those of lower ignition propensity.

The limit on the degree of resolution for this method is similar to that using mock-ups. Table 30 shows that the level of lab-to-lab reproducibility is about 0.4 for 16 replicate tests. Only modest improvement is achievable for a reasonably larger number of tests. It is apparent that the Cigarette Extinction Test also cannot make fine distinctions in ignition propensity among cigarette designs. *Again, this places a limit on the degree of resolution possible for regulatory use of the method.* As above, finer distinctions could be made within a single laboratory by using a number of replicates greater than 16.

It is also possible to calculate how one might use fewer tests on a substrate, having measured 100% ignitions on a substrate of higher thermal capacitance. For example, after 16 full-length burns in 16 tests on the 15-layer substrate, a 95% lower confidence bound on the long-run full-length burn probability is $(1-0.95)^{1/16} = 0.83$. In other words, one can be 95% confident that the full-length burn probability within the same laboratory is greater than about 0.83. With that result, it may be sufficient for some purposes to know that the full-length burn probability on the 10-layer and 3-layer substrates would be about 0.6 or higher, since the degree of interlaboratory resolution is about 0.4, as noted above. If so, then only about $n=6$ runs would be required, because $(1-0.95)^{1/6} = 0.61$. That is, if 6 runs were conducted resulting in 6 full-length burns, then the 95% lower confidence bound for the full-length burn probability would be 0.61. It is, of course, possible that in the 6 tests, one or more self-extinguishments might occur. It would then be appropriate to run a full set of 16 replications for each of the three substrates. Note that the savings in resources with this method is somewhat smaller than with the mock-up method.

C. Allowable Material Variability

The lab-to-lab reproducibility seen in each ILS for the two test methods (Figures 4 and 14) is a result of the variability of the test operators, the laboratory environment, the substrate materials, and the products being tested. Measuring variations in the cigarettes is part of the purpose of testing. Therefore, in order to assure that the test reproducibility is maintained at the observed level, the substrate material variability limits existent in the present study must be applied to all future materials. This may be a more stringent requirement than necessary but, without further study, there is no justification for looser controls on the materials.

For the Mock-Up Ignition Test Method, the most critical material is the fabric. Acceptable fabrics must be 100% raw cotton ducks which meet the physical requirements of reference 12.[8] Open-end spinning should also be specified to ensure similarity to the fabrics used in the interlaboratory studies. Since reference 12 does not explicitly specify such details as yarn count and yarn plies and the influence of these structural parameters on cigarette ignitability has not been extensively explored, the values listed in Table 6 should be adhered to. Where the limits on other properties are narrower for the cotton ducks actually used in this study, those narrower limits must apply. Thus the acceptable areal density and air permeability ranges are those given in Tables 11 and 12. Reference 12 contains no specification on metal ion content of the fabrics. The data in Table 7 for duck #6 suggest that potassium ion levels in the range from 4400 to 6000 ppm are acceptable. Comparison of the potassium levels in the various ducks in Tables 7 and 8 suggests that all of the ducks can generally be held in this range for an extended production period, though conclusions about multi-year variability obviously cannot be made on the basis of the present study. Sodium is potentially as catalytic to smoldering ignition as is potassium; thus it should (and probably naturally will) be held to the negligible levels seen in Table 7. Calcium and magnesium cations are weak smolder promoters; Table 7 suggests an acceptable range of both is 500 to 750 ppm. Cotton ducks should be stored in the dark at room temperature or below and at a humidity low enough to preclude any microbial action. Under these circumstances, shelf life should be at least one year.

As noted previously, the Mock-Up Method using duck #4 in combination with a polymer film as an added heat sink is sensitive to the properties of that film. Table 15 lists the properties of the Warp Brothers Poly-Film used in the main ILS. In reviewing the results shown in Figures 2 and 4, a film like this is preferred. The areal density is believed to be the most critical property here (along with the heat capacity of the film, which has not been measured but should be fixed by the composition). An areal density change from 0.015 to 0.012 g/cm^2 (Poly-America film) caused a substantial change in the ignition proportions of cigarettes 501 and 503. Cigarette 501, in particular, decreased from an ignition proportion range of 0.7-1.0 to 0-0.3 when the higher areal density film was used. This sensitivity suggests that the areal density should be held at 0.015 g/cm^2 ± 5 percent.

The polyurethane foam in the Mock-Up Method probably serves more of a physical role (part insulator, part heat sink, part oxygen inflow inhibitor) than a chemical role (source or sink of chemical energy). The foam does not smolder during the fabric ignition process or soon thereafter. In this study, substitution of another foam with a 25% lower density and a 17% greater air permeability had no great effect on the ignition proportions seen. Thus there should be no difficulties introduced by allowing the typical ± 5% within-batch variations in foam density and the level of air permeability variations seen in Table 14. Reference 9 indicates that there is a weak correlation between the number of urea bonds in the foam formulation and ignitability. The number of urea bonds is proportional to the water level in the foam formulation which is unknown for the foams in this study. The study reported in reference 9 also indicated that ignitability is an order of magnitude more sensitive to foam air permeability than to urea bond levels. Since these two foam properties are somewhat related (water is added as a foam blowing agent), the preceding restriction

---

[8] Reference 12 accepts recycled cotton as raw material. That is undesirable here because of the possibility of chemical contamination which could affect ignitability by cigarettes.

on foam density variations should suffice, provided the foam is a polyether/TDI formulation typical of current technology. The foam also should contain no inorganic fillers. Polyurethane foams show substantial color changes when exposed to typical room lighting. The possibility of significant alterations via this mechanism or via slow aging should be precluded by storage under an opaque covering at room temperature for no more than six months before use.

The filter paper is the only critical material used in the Cigarette Extinction Test Method. The paper used here is Whatman No. 2 filter paper, a staple for qualitative analysis. The nominal ash content specification is 0.06%. Here again the critical property is the areal density, since the principal function of the paper is to serve as a heat sink. It is likely that the thermal conductivity also plays a role here, especially in the substrates with the greatest number of paper sheets. This should be proportional to the density of the paper. The data in Section II.C.3 are for the filter papers used in the present study. The areal density and thickness determine the paper density as well. The variability of these properties shown there is acceptable.

Appendix F includes thermogravimetric data on all of the materials used in both test methods. These data are included as general guidance in assessing the suitability of candidate new batches of material. Any new batch of a given material should behave in substantially the same manner as the example in Appendix F for that type.

It is possible that a performance specification could be developed for mock-up materials based on a standardized, non-cigarette ignition source, such as a small black body or a carbon dioxide laser. This might obviate the need for all of the prescriptive limitations on materials given above. Other work in this program has shown that the smoldering ignition process is sensitive to such variables as the size of the spot on a mock-up surface which is heated. Thus any such performance test would require careful study of the sensitivity of the output (*e.g.*, ignition delay time for a range of peak incident heat fluxes) to test variables. Since the materials specifications above appeared to be sufficient, this alternative approach has not been pursued to completion.

D. Standardization of Test Methods

It is common practice, upon development of a fire test method for professional use, to proceed with its adoption as a voluntary consensus standard in either ASTM or the National Fire Protection Association (NFPA). Because these processes generally take several years, this is not possible under the Fire Safe Cigarette Act of 1990, which expires September 10, 1993. This report contains sufficient documentation of the two test methods and interlaboratory evaluations of each so that the methods could be submitted to ASTM or NFPA for formal approval. The relevance to fire safety is contained here and in reference 3. Thus, all necessary materials for initiating the standardization process are now available.

E. Effectiveness of the Methods

It will be the role of the regulator to determine which (if any) future cigarettes are tested, by whom, how frequently, and to what requirements. The last of these is likely to be based, in part, on the additional degree of life safety desired. The findings of the TSG demonstrated that measurements using mock-ups are reasonable indicators of full-scale performance [2]. The work to date provides modest guidance in relating performance under these new methods to real-world performance.

There are data to "calibrate" the methods at the high end of the ignition propensity scale. The current commercial cigarettes are associated with the fire losses of today. The commercial cigarette data in Section IV of this report and the data on older commercial cigarettes in reference 3 establish typical performance for these cigarettes. In the two new test methods, this is seen as a large number of ignitions on the #4 cotton duck or full-length burning on the 15-layer paper substrate. This establishes the test results for the high ignition propensity end of the scale.

Both the current work and cigarette industry studies [20] demonstrate the performance of cigarettes that never or rarely ignited a variety of substrates. The correlation of mock-up results with chair tests in reference 3 indicates that such results can be expected to be indicative of real-world performance of such substrates. In the two new test methods, this behavior is observed as few ignitions on the #10 cotton duck or few full-length burns on 3 layers of filter paper. This indicator of test results for the low ignition propensity end of the scale is less quantitative than the high end indicator mentioned earlier.

In between these extremes, one would like to be able to predict a reduced number of fires as fewer ignitions are measured in the laboratory. The full-scale tests in reference 3 support this. At least for coarse changes in test performance, real-world savings seem highly likely. When considering smaller increments in test performance, however, one must keep in mind the accuracy limits of the methods as discussed above.

IV. TESTING OF COMMERCIAL CIGARETTES

A. Introduction

Having completed the development of standardized testing procedures, NIST has evaluated a sample of current commercial cigarette types, its second obligation under the Fire Safe Cigarette Act of 1990. The results of this performance testing:

- demonstrate the utility of the method for routine testing of production cigarettes,

- provide baseline data for comparison with commercial cigarettes of the future, and

- present examples of the recommended reporting format for the ignition propensity data.

B. Rationale for Commercial Cigarette Choices

The cigarettes were chosen with two objectives in mind: (a) to incorporate packings which comprise a significant portion of the consumer market, and (b) to include several packings judged likely to yield a lower ignition propensity compared to the best sellers. After reviewing available physical characteristic data, fourteen packings in the former category were tested and six in the latter.

Consumer market data were obtained from the February 10, 1992 Maxwell Consumer Report. This includes complete sales data only through 1990, and it is the 1990 data which were used. The Maxwell Report data indicate that the fourteen packings chosen comprised 38% of the market in 1990.

These best selling brands do not vary widely in physical parameters such as packing density, paper permeability or tobacco rod circumference. Thus a second, smaller group of cigarettes was identified which do show more substantial deviations in their physical parameters. The particular emphasis was on cigarettes having two physical parameters which deviate in a direction which the TSG study [3] would suggest as likely to lower ignition propensity, *e.g.*, lower paper porosity, circumference, tobacco density. [Since the data on which these decisions were based were identified as confidential and proprietary by the cigarette industry, they are not tabulated in this report.] The six selected packings comprised less than one percent of the market in 1990.

Samples of these cigarettes were obtained courtesy of the Tobacco Institute Testing Laboratory (TITL), Rockville, Maryland. This laboratory is responsible for the official tar and nicotine ratings of all commercial cigarettes. It contracts to have packs of cigarettes purchased from around the country once each year, in the first quarter of the calendar year. The cigarettes tested in this study were from surplus packs purchased in the first quarter of 1992; typically the quantity obtained was 80-100 packs of each type. The cigarettes were in opened packs which had been sealed in plastic bags after TITL's sampling and stored at room temperature. They were also stored at NIST in this manner until removed for conditioning.

C. Test Procedures

The cigarettes were tested nominally using the two procedures described above. However, it was expected that many of these packings would be of very high ignition propensity. Therefore, the full complement of replicates (48 or 16) was performed first for the duck #4 and 15-layer substrates. If 48 ignitions in the Mock-Up Ignition Method were observed, then 8 replicates were performed on the remaining substrates. (This is more than the minimum of six called for in Section II.A., since it is based on an earlier estimate of the reproducibility of the test methods.) Also, for certain of the top fourteen packings (4 through 9 in Table 32), additional tests were run on Duck #6 before the decision was made to limit the testing to eight replicates. If 16 full-length burns in the Extinction Method were noted, then no further testing was performed.

One of the packings in the group of six selected as less ignition prone (packing C in Table 32) had a tendency to self-extinguish during the vertical free-burn period prior to placement on the substrate. Thus, on duck #6, 13 of the 35 extinguishments occurred during this vertical free-burn. When an additional 24 replicates were run with this duck using a horizontal cigarette orientation during the free-burn interval, 12 caused ignition and 12 self-extinguished on the mock-up. This increase in ignition rate from 27% to 50% is comparable to the repeatability of the method (Table 24) and was not considered sufficient to revise the test procedure. Note, however, that this cigarette was run only with a horizontal free-burn for the tests on duck #10 and the three filter paper substrates.

D. Analysis of Data

Table 32 shows that the top 14 best-selling packings behaved in a virtually identical manner, with packing #7 exhibiting 2 self-extinctions on the duck #4. Both test methods indicate they are strong igniters; neither method reveals any differentiation among these packings. Reference to the interlaboratory study results for the Mock-Up Ignition Method (Figure 4) indicates that all of these cigarettes are stronger igniters than the 2 strongest experimental cigarettes (503 and 501) used in that study.

The 6 packings chosen as likely to be of lesser ignition propensity did in fact show this tendency to varying degrees. Both methods reveal the same qualitative picture: a monotonic increase in the number of ignitions or full-length burns as one moves toward the lighter fabrics or fewer filter layers. Of particular note is that 4 of these packings (A, C, E, F) showed few or no ignitions in 48 replicates on the #4 duck. Compared to the Mock-Up Ignition Method, the Extinction Method does not seem to pick up the reduced ignition propensity of one of these (packing E) and also does not distinguish as strongly the performance of cigarettes A and D from the 14 best-selling packings. Packing C shows a persisting tendency toward a lesser ignition propensity, even on Duck #10; the Extinction Method does not show this on the 10- or 3-layer substrates. These observations are consistent with those in the interlaboratory study, which indicated that the Mock-Up Method is capable of better distinction among cigarettes in the upper/middle part of the ignition propensity range.

Table 32. Results of Commercial Cigarette Testing

Cigarette	Mock-Up Ignition Test Method			Cigarette Extinction Test Method		
	Duck #4[9]	Duck #6	Duck #10	15 Layers[10]	10 Layers	3 Layers
1	48/0/0	8/0/0	8/0/0	16/0		
2	48/0/0	8/0/0	8/0/0	16/0		
3	48/0/0	8/0/0	8/0/0	16/0		
4	48/0/0	12/0/0	8/0/0	16/0		
5	48/0/0	12/0/0	8/0/0	16/0		
6	48/0/0	12/0/0	8/0/0	16/0		
7	46/0/2	16/0/0	8/0/0	16/0		
8	48/0/0	16/0/0	8/0/0	16/0		
9	48/0/0	16/0/0	8/0/0	16/0		
10	48/0/0	8/0/0	8/0/0	16/0		
11	48/0/0	8/0/0	8/0/0	16/0		
12	48/0/0	8/0/0	8/0/0	16/0		
13	48/0/0	8/0/0	8/0/0	16/0		
14	48/0/0	8/0/0	8/0/0	16/0		
A	2/5/41	44/0/4	8/0/0	6/10	15/1	6/0
B	35/1/12	44/0/4	8/0/0	15/1	6/0	6/0
C	0/0/48	13/0/35	4/0/4	2/14	8/8	16/0
D	22/1/25	35/0/13	8/0/0	14/2	15/1	6/0
E	0/31/17	46/0/2	8/0/0	15/1	16/0	6/0
F	0/6/42	38/0/10	8/0/0	3/13	8/8	16/0

Table 33 provides a further check of consistency between the two methods, and thus further affirmation that the measured cigarette performance is consistent across diverse substrates. Here, the ignition strengths of the five cigarettes from the interlaboratory study and five of the second group of commercial cigarettes are tabulated. [The fourteen best-selling commercial packings showed nominally 100% ignitions on all six substrates and thus the data are not informative.] Cigarette C is omitted because the testing was performed using two different pre-burn procedures. The rows are in order of decreasing average of the six values in the row; the columns are arranged similarly. The averages from the interlaboratory study are for 432 replicates (9 labs × 48 each) for the cotton duck substrates and 144 replicates (9 × 16) for the filter paper substrates. The number of replicates for the commercial cigarettes are far fewer and shown in Table 32.

---

[9] Results for the Mock-Up Ignition Method are shown in the sequence: Ignitions/Non-Ignitions/Self-Extinctions.

[10] Results for the Cigarette Extinction Method are shown in the order: Full Burn/Partial Burn.

Table 33. Percent Ignitions or Full Length Burns on Test Method Substrates

SUBSTRATE → CIGARETTE ↓	3 Layers	Duck #10	10 Layers	Duck #6	15 Layers	Duck #4
B	100	100	100	92	94	73
503	100	100	100	100	100	53
501	100	100	100	100	100	11
D	100	100	94	73	88	46
E	100	100	100	96	94	0
531	99	98	94	95	88	0
A	100	100	94	92	38	4
F	100	100	100	79	19	0
529	57	30	6	8	2	0
530	6	3	0	0	0	0

There is a generally consistent decrease in ignition strength from the top left corner of the matrix to the lower right corner, especially considering the reproducibility of the data established in the interlaboratory study. Perhaps the largest single departure from the general pattern in the Table is for either cigarette 501 or cigarette D tested on the duck #4 substrate. However, these cigarette ignition propensities are quite comparable to each other. The two duck #4 values are within the established interlaboratory reproducibility of each other, and both cigarettes yield similar results on all the other substrates.

As in the TSG studies [3], it is of interest to determine whether reduced ignition propensity necessarily results in increased yields of undesirable smoke components. The mean values and standard deviations for the two sets of commercial cigarettes tested here are shown in Table 34. The entries in Table 34 were compiled from data contained in reference [34]. These data were generated by the Tobacco Institute Testing Laboratory. The results show that reduced ignition propensity has been achieved with no significant increase in these three smoke components.

Table 34. Averaged Smoke Component Yields from Commercial Cigarettes (mg per cigarette)

Cigarettes	Tar (mg)	Nicotine (mg)	Carbon Monoxide (mg)
1-14	14.4 ± 4.2	1.04 ± 0.27	13.7 ± 2.2
A-F	11.7 ± 4.8	0.98 ± 0.38	12.5 ± 6.2

V. CONCLUSIONS AND RECOMMENDATIONS

The research funded under the Cigarette Safety Act of 1984 (P.L.98-567) and the Fire Safe Cigarette Act of 1990 (P.L. 101-352) has led to the development of two test methods for measuring the ignition propensity of cigarettes.

- The Mock-Up Ignition Test Method uses substrates physically similar to upholstered furniture and mattresses: a layer of fabric over padding. The measure of cigarette performance is ignition or non-ignition of the substrate.

- The Cigarette Extinction Test Method replaces the fabric/padding assembly with multiple layers of common filter paper. The measure of performance is full-length burning or self-extinguishment of the cigarette.

The 14 best-selling commercial cigarette packings and six other commercial packings were examined using the two methods. Both methods showed reduced ignition propensities for five of the six specialty cigarettes relative to the best sellers.

For a product standard at present, there is a preference for using the Mock-Up Ignition Test Method because it is capable of better discrimination among cigarettes of high/moderate ignition propensity. However, routine measurement of the relative ignition propensity of cigarettes is feasible using either of the two methods.

Improved cigarette performance under both methods has been linked with reduced ignition behavior in full-scale chairs constructed using fabrics that differ substantially from the materials in the test methods. It is reasonable to assume that this implies an analogous benefit of reduced ignitability is to be found in the real world population of upholstered furniture. However, the *precise* incremental life and property savings that would accrue from the use of the test methods described here in conjunction with a particular test criterion has not been established.

Both methods have been subjected to interlaboratory study. The resulting reproducibilities were comparable to each other and comparable or superior to most currently-used standard fire tests.

VI. ACKNOWLEDGMENTS

The authors gratefully acknowledge the helpful assistance of numerous persons who helped make this study possible. The Consumer Product Safety Commission funded this program. Throughout this study, there were many helpful interactions with CPSC staff, especially Bea Harwood, James Hoebel, and Roy Deppa, who also served as technical monitor for the NIST work.

Michael Smith and Hartley Abraham performed the majority of the cigarette testing with assistance from Jack Lee, Russell Loy, Arnold Liu and Ronald McCombs. Henry Wheelock designed and fabricated much of the test apparatus. John Shields provided extensive help in efforts to prepare doped fabrics and in characterization of the ignition process. Robin Breese assisted in thermal analyses and plotting the data. Dick Zile, Lauren DeLauter, Gary Roadarmel and Roy McLane helped with the randomizing of the materials samples. John Krasny provided helpful discussions throughout the study.

Lisa Oakley, of the NIST Statistical Engineering Division, assisted with statistical computing and preparation of statistical graphics for this report; Dr. Raghu Kacker, also of the NIST Statistical Engineering Division, participated, along with Magdalena Navarro, in designing the interlaboratory studies and the NIST experimental work that preceded them; and Dr. Terry Kissinger of the Consumer Product Safety Commission provided helpful suggestions on the statistical analyses.

Various people in industry provided useful information in specific areas. Dr. Alexander Spears of Lorillard coordinated the cigarette industry manufacture of the Series 500 experimental cigarettes and also the transferring of the commercial cigarettes from the Tobacco Institute Testing Laboratory to NIST. Wayne Smith of West Point Pepperell was especially helpful in providing data on the potassium content of cotton ducks. Hardy Poole of the American Textile Manufacturers Institute provided alternative cotton ducks for assessment. Paul Weinle, Hugh Talley, and Bill Jones, Mike McDonnell, Joe Guerriera and many others who provided property information on materials are gratefully acknowledged.

Finally, the authors wish to thank the members of the Technical Advisory Group under the Fire Safe Cigarette Act of 1990 for helpful comments on this text and to the staffs of the nine laboratories that participated in the interlaboratory study.

VII. REFERENCES

[1] Miller, A.L., "U.S. Smoking Material Fire Problem Through 1990: The Role of Lighted Tobacco Products in Fire," National Fire Protection Association, Quincy MA, 1993.

[2] "Toward a Less Fire-Prone Cigarette," Final Report to the Congress, Technical Study Group on Cigarette and Little Cigar Fire Safety, Cigarette Safety Act of 1984, available from the U.S. Consumer Product Safety Commission, 1987.

[3] Gann, R.G., Harris, Jr., R.H., Krasny, J.F., Levine, R.S., Mitler, H.E., and Ohlemiller, T.J., "The Effect of Cigarette Characteristics on the Ignition of Soft Furnishings," NBS Technical Note 1241, National Bureau of Standards, Gaithersburg, MD, 1987.

[4] Ihrig, A., Rhyne, A., Norman, V. and Spears, A., "Factors Involved in the Ignition of Cellulosic Upholstery Fabrics by Cigarettes," *J. Fire Sciences* **4**, 237 (1986).

[5] California Bureau of Home Furnishings and Thermal Insulation, Technical Bulletin 117. "Requirements, Test Procedure and Apparatus for Testing the Flame Retardance of Resilient Filling Materials Used in Upholstered Furniture."

[6] Harris, Jr., R.H., Navarro, M., Gann, R.G., and Eberhardt, K.R., "Reevaluation of Experimental Cigarettes used in the Cigarette Safety Act of 1984," NIST Report of Test FR 3984, National Institute of Standards and Technology, Gaithersburg, MD, 1991.

[7] Babrauskas, V. and Krasny, J., "Fire Behavior of Upholstered Furniture," National Bureau of Standards Monograph 173, 1985.

[8] "Cigarette Ignition Resistance of Components of Upholstered Furniture," NFPA 260-1989, National Fire Protection Association, Quincy, MA, published annually.

[9] Ihrig, A., Rhyne, A. and Spears, A., "Influence of Flexible Polyurethane Foams on Upholstery Fabric-Foam Mock-ups Ignitions by Cigarettes", *J. Fire Sciences* **5**, Nov/Dec, 1987, p. 392.

[10] Rhyne, A. and Spears A., "Sensitivity of a Cigarette Ignitability Index to Hypothetical Shifts in the Distribution of Upholstery Furniture Fabrics", *J. Fire Sciences* **8**, Jan/Feb, 1989, p. 3.

[11] McCarter, R., "Smoldering Combustion of Cotton and Rayon," *J. Consumer Prod. Flamm.* **4**, 1977, p. 346.

[12] CCC-C-419G, December 15, 1989 "Federal Specification - Cloth, Duck, Unbleached, Plied-Yarns, Army & Numbered," available from U.S. Army Natick Research, Development and Engineering Center, Natick, MA 01760-5014.

[13] Donaldson, D., USDA Southern Regional Research Laboratory, personal communication.

[14] Mauersberger, H., ed., <u>Mathews Textile Fibers, Sixth Edition</u>, J. Wiley & Sons, New York, (1954), p.225.

[15] An informal sampling of references to the use of cotton ducks and duck-like fabrics identified the following: IKEA, 1992 catalog; The Crate and Barrel, 1992 Spring and Summer catalog; Marlo Furniture Inc., 1992 spring sale literature; Ethan Allen Inc., spring sale event literature; Metropolitan Home magazine, April, 1992, p. 136; Better Homes and Gardens magazine, March, 1993, p. 131.

[16] Norman, V., Lorillard, Inc., presentation to the Technical Advisory Group, Fire Safe Cigarette Act of 1990, April 16, 1992.

[17] ASTM Method D 737-75, "Standard Test Method for Air Permeability of Textile Fabrics," American Society for Testing and Materials, Philadelphia, PA, 1980.

[18] ASTM Method D 3574-91, "Standard Test Methods for Flexible Cellular Materials--Slab, Bonded, and Molded Urethane Foams," American Society for Testing and Materials, Philadelphia, PA.

[19] Dowling, V., Division of Building, Construction and Engineering, CSIRO, Australia, personal communication.

[20] Adiga, K., Pham, M., Noonan, K. and Honeycutt, R., "The Implications of Modest Air Flow on Cigarette Ignition of Soft Furnishing Mockups," report submitted to the Technical Advisory Group on October 15, 1992 by Brown and Williamson Tobacco Corporation.

[21] Countryman, C., "Some Physical Characteristics of Cigarettes as Fire Brands," Report prepared under U. S. Forest Service Contract 43-9AD6-1-617 (1981).

[22] Flack, R., University of Virginia, report to the Technical Advisory Group, January 28, 1993.

[23] Robinson, D., "Aerodynamic Characteristics of the Plume Generated by a Burning Cigarette," Paper presented at the International Conference on the Physical and Chemical Processes Occurring in a Burning Cigarette, sponsored by R. J. Reynolds Tobacco Co., held at Wake Forest University, April, 1987.

[24] ASTM E 177-90a, "Standard Practice for Use of Terms Precision and Bias in ASTM Test Methods," American Society for Testing and Materials, Philadelphia, PA 1990.

[25] ASTM E 691-87, "Standard Practice for Conducting an Interlaboratory Study to Determine the Precision of a Test Method," American Society for Testing and Materials, Philadelphia, PA, 1988.

[26] Cox, D.R., and Snell, E.J., *Analysis of Binary Data*, 2nd edition, London: Chapman and Hall, 1989, section 3.2.

[27] D. A. Williams, "Extra-Binomial Variation in Linear Logistic Models," *Applied Statistics*, Vol. 31 (1982), No. 2, pp. 144-148.

[28] Finney, D. J., *Probit Analysis*, 3rd Edition, London, Cambridge University Press, 1971.

[29] ASTM Method E 648-91, "Standard Method for Critical Radiant Flux of Floor-Covering Systems," American Society for Testing and Materials, Philadelphia, PA.

[30] E 662-83, "Standard Test Method for Specific Optical Density of Smoke Generated by Solid Materials," American Society for Testing and Materials, Philadelphia, PA.

[31] E 1354-90, "Standard Test Method for Heat and Visible Smoke Release for Materials and Products Using an Oxygen Consumption Calorimeter," American Society for Testing and Materials, Philadelphia, PA.

[32] Krasny, J., Allen, P., Maldonado, A. and Juarez, N., "Development of a Candidate Test Method for the Measurement of the Propensity of Cigarettes to Cause Smoldering Ignition of Upholstered Furniture and Mattresses," National Bureau of Standards, NBSIR 81-2363, 1981.

[33] Norman, V., "Some Considerations Concerning the Heat Output of Cigarettes," Presentation to the Technical Study Group, March, 1985.

[34] "Tar, Nicotine, and Carbon Monoxide of the Smoke of 534 Varieties of Domestic Cigarettes," Federal Trade Commission, Washington, DC, 1992.

APPENDIX A

U.S. DEPARTMENT OF COMMERCE
NATIONAL INSTITUTE OF STANDARDS AND TECHNOLOGY
Gaithersburg, MD 20899

Report of Test
FR 3984

May 1991

"Reevaluation of Experimental Cigarettes used in the Cigarette Safety Act of 1984"

Richard H. Harris, Jr., Magdalena Navarro, Richard G. Gann

Building and Fire Research Laboratory
National Institute of Standards and Technology
Gaithersburg, MD 20899

Keith R. Eberhardt

Computing and Applied Mathematics Laboratory
National Institute of Standards and Technology
Gaithersburg, MD 20899

Submitted to:

U.S. Consumer Product Safety Commission
Washington DC 20207

Introduction

Under P.L. 101-352, the Fire-Safe Cigarette Act of 1990, one of the tasks assigned to NIST was to develop a valid test method for the ignition propensity of cigarettes. In a previous study [A-1], under the Cigarette Safety Act of 1984, 32 experimental cigarettes that vary systematically in five design parameters were extensively studied to determine their ignition propensity on soft furnishings. The evaluation was done with bench-scale mock-ups and on full-scale furniture. Analysis of the test results indicated an agreement between the bench-scale and full-scale tests. Since these experimental cigarettes are well characterized, it is desirable to use these cigarettes in the new study. However, these cigarettes have been stored in freezers since the completion of the first study. To determine if the cigarettes have changed during storage, a reevaluation was undertaken using fabric and padding materials retained from the first study.

Statistical Selection of Cigarettes

The original characterization of these cigarettes involved testing on four substrates, which in turn were composed of combinations of 3 fabrics and two paddings. Due to the limited availability of one of the fabrics (California standard), it was impossible to repeat the entire experimental series using all 32 types of cigarettes on all 4 substrates. There was only enough material available to perform ignition tests for 8 of the 32 cigarettes on all 4 substrates, using 5 replicates for each case.

The selection of 8 cigarettes from the available 32 amounts to choosing exactly a 1/4 fraction of the available cigarette types. In making the selection, we attempted to achieve two objectives:

(1) to choose cigarettes whose ignition propensities evenly span the entire range of ignition rates observed in the previous testing, and

(2) to choose cigarettes in a balanced fashion — so that each of the five design factors that define the cigarette types would be equally represented among the 8 selected cigarettes.

The statistical theory of *fractional factorial* experimental design [A-2] can be used to satisfy the second objective. In particular, a 1/4 fraction of the (full) 2^5 factorial design that defines the 32 cigarettes would consist of 8 cigarettes for which:

- 4 have Burley tobacco, and 4 have Flue-cured;
- 4 have Expanded tobacco, and 4 are Not expanded;
- 4 have paper of Low permeability, and 4 have High permeability;
- 4 have Citrate, and 4 have No citrate; and
- 4 are 21 cm in circumference, and 4 are 25 cm.

In addition, in a fractional factorial experimental design, a second level of balance would be achieved. For example, among the 4 cigarettes with high citrate, 2 would be 21 cm in circumference and 2 would be 25 cm; similarly, the 4 low citrate cigarettes would have 2 at 21 cm and 2 at 25 cm circumference. In an analogous way, each pair of factors would exhibit this kind of balance, with the result that each level of one factor would be combined with each level of the other factor in an equal number of cases.

There are many ways that this kind of fully balanced fractional factorial selection could be made from the 32 cigarette types available. It was initially hoped that one or more of these fractional factorial selections would yield a set of 8 cigarettes that would also satisfy the first objective of uniformly spanning the ignition rates that had been obtained in the previous experiment. Ultimately we found that it was not possible to achieve both of the stated objectives exactly, and so a compromise set of 8 cigarettes was found that was imperfectly, but nearly, balanced and which does exhibit quite uniform coverage of the ignition rates. It was felt that for the purposes of this reevaluation experiment, the need to use cigarettes that uniformly represent the full range of previously observed ignition rates was more important than achieving a perfectly balanced fractional factorial arrangement.

Table A-1 displays the extent to which balance in the above sense was achieved in the final compromise set of cigarettes chosen.

Table A-1. Selection of Cigarettes for Reevaluation Study: Balance on Cigarette Design Factors and Coverage of Levels of Previous Numbers of Ignitions

Cigarette Number	Tobacco Type	Packing Density	Paper Permeability	Citrate	Circumference	Previous Number of Ignitions
106	B	E	L	N	21	1
130	F	E	L	N	25	4
108	B	E	H	N	21	7
129	F	E	L	C	25	10
101	B	N	L	C	21	13
131	F	E	H	C	25	15
103	B	N	H	C	21	17
120	B	N	H	N	25	20
Balance Achieved	5 B 3 F	5 E 3 N	4 L 4 H	4 N 4 C	4 21mm 4 25mm	

Results and Conclusions

The eight statistically-selected cigarettes were tested for their ignition propensity on the same substrates and in the same manner as the previous study. In addition to the storage factor, two other differences were a change in the canopy hood used and the technician who performed the tests.

The results of the testing are shown in Table A-2 below.

Table A-2. Reevaluation of Eight of the Thirty-two Experimental Cigarettes

Cig. Id.	Number of Ignitions									
	CA/CB Unc./flat		SPL/PU Unc./flat		SPL/PU[a] Unc./flat		Denim/PU Crev./cov.		Total	
	Prev.	Now	Prev.	Now	Prev.	Now	Prev.	Now	Prev.	Now
106 BELN21	0	0	1	1	0	0	0	0	1	1
130 FELN25	3	4	1	1	0	2	0	5	4	12
108 BEHN21	3	2	4	3	0	0	0	1	7	6
129 FELC25	5	5	3	5	2	5	0	1	10	16
101 BNLC21	3	4	5	5	5	5	0	5	13	19
131 FEHC25	5	3	5	5	5	5	0	0	15	13
103 BNHC21	5	4	5	5	5	5	2	5	17	19
120 BNHN25	5	5	5	5	5	5	5	5	20	20
Totals	29	27	29	30	22	27	7	22	87	106

Maximum number of ignitions per cigarette is 20, per substrate 40.
CA/CB California test fabric/cotton batting.
SPL/PU 100% cotton Splendor fabric/polyurethane 2045.
Denim/PU 100% cotton Denim fabric/polyurethane 2045.
Unc./flat Uncovered cigarette on a flat mockup.
Crev./cov. Covered cigarette in mockup crevice.
a Cigarette with filter with one half of the tobacco column removed before lighting.

If no real change in ignition propensity occurred, then the numbers of ignitions in the "Previous" and "Now" columns of Table A-2 should be the same, except for statistical fluctuations.

For each of the 8 cigarette types, and for each of the 4 substrates shown in Table A-2, we calculated the difference between the number of ignitions in the current study ("Now"), minus the number of ignitions in the "Previous" study. If these differences represent only statistical fluctuations, then they would form a statistical population centered near zero. The Wilcoxon Signed Rank Test [A-3] was adopted as a formal statistical test procedure to evaluate whether the observed differences indicate a change in ignition propensity or only random noise. This is a non-parametric test procedure that is valid for use with data that do not follow the commonly assumed Gaussian distribution. Validity for non-Gaussian data was an important consideration because the difference data from this experiment clearly exhibit a non-Gaussian pattern of variation.

The results of the Wilcoxon Signed Rank Test are that the observed differences in numbers of ignitions show a statistically significant tendency ($p = 0.04$) toward increased ignitions after the storage period. Inspection of Table A-2 shows that the increased ignitions come almost exclusively from the denim substrate, which suggests the possibility that the statistically significant difference is due entirely to the denim substrate. This is consistent with the observation of Rogers and Hayes [A-4] that unless denim is stored free of finishing materials in the dark and in a temperature controlled environment, it will deteriorate with time.

To evaluate the hypothesis of no change in ignition propensity for the non-denim substrates, the Wilcoxon Signed Rank Test was recomputed using only the other three substrates (CA/CB, SPL/PU, and SPL/PU-half cigarette). In this case, the differences in ignition numbers were not significantly different from zero ($p = 0.47$). That is, the data for the three non-denim substrates are wholly consistent with the hypothesis of no change in ignition propensities of the experimental cigarettes, compared with the previous study.

It was noted that cigarette number 129 showed noticeable increases in the number of ignitions for both of the conditions involving the SPL/PU substrate. This suggests the possibility of a real change in ignition propensity for this particular combination of cigarette and substrate. In pursuing this observation, it is pertinent to note that cigarette 129 showed a relatively small increase in ignitions on the denim substrate both in comparison to its increase for the two SPL/PU substrates and also in comparison to the increases for other cigarettes on the denim substrate. Thus any physical explanation of a change in ignition propensity for cigarette 129 would seem to call for a unique cigarette-substrate interaction on SPL/PU. The reevaluation experiment was designed as an overall test for possible changes in the experimental cigarettes. It was not designed to generate sufficient data to evaluate unique effects for each cigarette and substrate combination. As it happens, the largest single observed difference (2 ignitions in the previous study versus 5 ignitions now) is not significant at the standard 5% level of significance ($p = 0.08$). Here, the significance calculation was obtained using Fisher's Exact Test for a 2×2 contingency table [A-5]. Based on all these considerations, it does not seem profitable to pursue further the observed increase in ignitions for cigarette 129 on SPL/PU.

Summary

Overall, we interpret the results of this study as showing that the ignition propensities of the experimental cigarettes have not changed during storage but that the denim substrate has changed in ignitability.

Acknowledgements

The authors thank Rajauhn Lee for performing the cigarette ignition tests for this project.

References

[A-1] Gann, R.G., Harris, R.H., Krasny, J.F., Levine, R.S., Mitler, H.E., and Ohlemiller, T.J., "The Effect of Cigarette Characteristics on the Ignition of Soft Furnishings," NBS Technical Note 1241, National Institute of Standards and Technology, Gaithersburg, MD, 1988.

[A-2] Box, G.E.P., Hunter, W.G., and Hunter, J. S., *Statistics for Experimenters*, John Wiley and Sons, New York, 1978, Chapter 12.

[A-3] Snedecor, G.W. and Cochran, W.G., *Statistical Methods*, 7th ed., The Iowa State University Press, Ames, Iowa, 1980, pp. 141-143.

[A-4] Rogers, R.E. and Hays, M., "Effect of Storage on Fabrics," *Textile Research*, pp. 22-35, April 1943.

[A-5] Placket, R.L., *The Analysis of Categorical Data*, 2nd ed., Macmillan, New York, 1981, Section 6.3.

APPENDIX B

MOCK-UP IGNITION TEST METHOD PROCEDURE

Scope:

The purpose of this test procedure is to provide a method for measuring the propensity of cigarettes to ignite specified types of substrate assemblies.

Apparatus and Equipment:

1. Test chambers of the design shown in Figure B-1 shall be constructed for testing the cigarette/substrate combinations.

2. A vacuum draw apparatus consisting of the following type shall be used for igniting the test cigarettes: See Figure B-2.

   > The vacuum draw apparatus is composed of a cigarette holder, particulate filter, rotameter, trap, pressure gauge, shut-off valve and a vacuum pump. The cigarette holder (a plastic drying tube) shall have a flexible diaphragm with a hole cut to a size appropriate for holding a cigarette and making a tight seal around the filter end. The holder must be mounted in a secure fastener which will hold the cigarette firmly in a horizontal position for lighting. The particulate filter, composed of a plastic drying tube 150 mm long and 20 mm in diameter filled with glass wool, shall be adequate to remove smoke from the combustion gases to prevent contamination of all downstream assemblies. The filter shall be changed regularly to insure that gas flow is unobstructed and that contamination is not allowed to pass this point. A rotameter capable of measuring 1000 ml/min, air, shall be used for adjusting flow through the vacuum draw system. A 250 ml impinger trap is used in the system to dampen flow variations. A vacuum gauge capable of measuring to at least 800 mm Hg shall be positioned before the pump in order to track apparatus operating pressure. The electric pump shall be capable of producing a vacuum of 500 mm Hg with a flow rate of at least 23 l/min. A Cole-Parmer Air Cadet pump model 7530-40 or equivalent is found to be suitable.

3. An environmental conditioning room or chamber shall be maintained which provides area adequate for conditioning cigarettes and test substrate materials. This room shall be capable of maintaining a relative humidity of 55 ± 5 % and a temperature of 23 ± 3 °C.

4. A constant humidity box of a size to hold 4 to 6 substrate assemblies and more than one 250 ml beaker of conditioned cigarettes is necessary in test rooms where humidity and temperature control is difficult. A shallow tray with a 15 mm deep layer of saturated solution of sodium bisulfate and water has been found to provide the appropriate conditioning environment. (NOTE: This solution is highly corrosive.)

5. The conditioning room and test rooms shall be monitored by a recording hygrothermograph (Cole-Parmer Model 8368-00 or equivalent). A Vista Scientific Corporation, battery operated psychrometer or equivalent is used to measure relative humidity in the constant humidity box and to calibrate the hygrothermograph.

6. Chemical or canopy hoods are needed for removing combustion products from the test room. Air flows through these hoods shall be at a level which is necessary to remove cigarette and substrate combustion products while not being high enough to influence combustion processes in the test chambers.

7. Clean plastic or rubber gloves shall be provided for test operators when handling test substrate materials and cigarettes.

8. A laboratory balance capable of weighing to 0.001 g with a repeatability of ± 0.003 g is required for weighing cigarette specimens.

9. A butane gas lighter capable of producing a stable luminous flame no longer than 20 mm is required for igniting test cigarettes.

10. Water spray bottles or 25 mm diameter wax candles may be used for extinguishing smoldering substrate assemblies. Appropriate fire-proof waste containers shall be used for disposal of the ignition test materials.

Calibrations:

The following are guidelines for basic calibrations of test equipment used in this standard. Time intervals stated in this method for calibrations are considered to be the minimum. Calibrations of equipment shall be carried out at any time when equipment or test conditions indicate that evaluation and recalibration may be necessary.

1. The ignition test chambers shall be checked before use to insure that the front door seals properly and that air movement in the test area does not introduce transient air movement in the test chambers. Door seals are checked visually to insure that they are closed flush against the chamber's side wall and the latching device secures the door tightly. All construction seams shall be inspected to insure that they are air tight and no cracks shall be visible on any surface of the test chamber. If leaks are detected, measures shall be taken to insure that these areas are again made air tight.

    Stability of air inside of test chamber shall be determined by making a substrate mock-up and placing it and a lighted cigarette on it in the test position. Observe air movement in the chamber to insure that smoke being emitted by the cigarette is rising vertically and is not showing turbulence within 150 mm above the lit end of the cigarette. If turbulence is noted: 1) the test chamber shall be checked for leaks, 2) the test chamber location shall be evaluated for excess air flow in the laboratory, and 3) air flow rate of the exhaust system shall be evaluated as the source of disturbance. Air flow in the test chamber shall be maintained to produce a near laminar vertical smoke stream to a minimum distance of 150 mm above the cigarette.

2. The vacuum draw apparatus for igniting cigarettes shall be calibrated each week before the beginning of testing. A soap film, bubble flow meter shall be connected to the inlet of the vacuum apparatus at the cigarette attachment point. The vacuum pump is started and adjusted to provide an air flow of 1000 ± 50 ml/min. This flow and range of variation shall be noted and recorded in the laboratory's calibration records for the vacuum system's rotameter.

3. The humidity and temperature sensors used to record environmental conditions in the conditioning room/chamber and test room shall be checked for accuracy each week. This shall be accomplished by comparing their results with a calibrated psychrometer and thermometer. A record of calibration adjustments shall be kept for each hygrothermograph used in the conditioning and test rooms.

4. The laboratory balance shall be calibrated each week. The calibration is checked by leveling the balance and weighing a calibrated 1.00 g mass. If the balance is found not to be within specifications, make adjustments as needed to bring the device into compliance. Record the results in your laboratory's calibration files.

Test Specimens and Substrates:

Cigarette test specimens and the test substrates are sensitive to contamination. At all times when these materials are handled, clean plastic or rubber gloves shall be worn.

1. Cigarette test specimens shall be protected from physical or environmental damage while in storage. It is important that the specimens not be crushed or deformed in any manner. Measures shall be taken to insure that the specimens are not contaminated by foreign materials while in storage and they shall be protected from degradation by insects. If the specimens are to be stored for more than one week, they shall be placed in a freezer reserved for the sole protection of the cigarette specimens.

2. Suitable constituent materials for the mock-ups are only those specified in the main body of this report. Bulk materials must be cut to size and then randomized in such a manner as to ensure that samples from any part of the material batch are equally likely to be incorporated in any given mock-up assembly.

3. Substrate materials consist of 200 x 200 mm (8 x 8 in) cotton duck fabric swatches and 200 x 200 x 50 mm (8 x 8 x 2 in) polyurethane foam cushions. The substrates are formed by squarely laying the cotton duck fabric flat on top of the foam cushion. Substrate construction may vary by placing various membranes or films, of equal dimensions as the fabric (except thickness), squarely between the cotton duck and foam cushion. All wrinkles shall be smoothed to produce a level test assembly with good contact between the layers.

Conditioning:

1. Cigarettes and substrate materials are conditioned at 55 ± 5% RH and 23 ± 3 °C for 24 hours prior to ignition testing. While conditioning, the cigarettes are contained vertically, with filter up, in clean 250 ml polyethylene or glass beakers with a maximum of 20 cigarettes per beaker to assure free air access to the specimens.

2. The substrates, composed of fabric, foam and possibly film materials, are positioned to allow air to circulate around the sides and top of the material. The substrates shall be placed on clean, dry surfaces while being conditioned.

3. If the laboratory conditioning room cannot meet the required environmental conditions, a controlled humidity box may be used. Humidity is maintained by the addition of water and chemicals to a holding tray located inside the box. Air in the box must be circulated by means of a small fan in one corner. If a relative humidity box is required, the RH and temperature must be measured with a calibrated wet/dry bulb hygrometer. A battery-driven fan instrument such as a Vista Scientific Corp. psychrometer, or equivalent, has been found to be suitable. Humidity measurements are made in the morning and evening of test day and recorded.

Safety:

1. Exhaust systems should be checked daily to insure that they are working. All products of combustion should be removed from the laboratory work area.

2. Personnel shall be instructed on general emergency procedures in the laboratory and on procedures to handle an uncontrolled fire.

3. Appropriate fire extinguishment equipment shall be provided in each fire test laboratory to extinguish test specimens and to suppress a small fire which may exceed normal controlled limits.

4. An appropriate waste container shall be used in each fire test laboratory for safe disposal of specimens and test assemblies after being exposed to heat and fire.

Test Procedure:

1. Turn on the exhaust system designated for removal of test combustion products 30 min prior to beginning ignition testing.

2. Do not start additional test preparations until relative humidity and temperature in the test room are stabilized within the following respective ranges 55 ± 5 % RH and 23 ± 3 °C. Record the relative humidity and temperature in your laboratory log.

3. Wear clean, plastic or rubber gloves when handling cigarettes and substrates.

4. Adjust the ignition source, a disposable butane gas lighter, to provide a flame no less than 15 mm in length and no longer than 20 mm.

5. If the testing room is separated from the conditioning room, individual substrate assemblies shall be placed into plastic bags and sealed in the conditioning room. They may then be transported to the test room. Test cigarettes may also be transported to the test room in sealed plastic bags.

    If the relative humidity and temperature in the test room cannot be maintained at the specified test conditions, the substrates shall be placed in the above mentioned humidity box prior to testing.

6. Immediately before selection and ignition of a cigarette for testing, a substrate assembly is removed from its conditioned environment and placed inside the test chamber at the geometric center of its bottom. The square brass frame is placed onto the substrate to maintain fabric flatness. Make sure that the fabric is flat against the foam surface by lightly smoothing with a gloved finger. Do not use fabrics that will not lay flat against the foam.

7. Without delay, remove a cigarette from the humidity box. Weigh the cigarette and record the results. Discard the specimen if its weight is more than two standard deviations of the mean obtained from weighing 50 random selected cigarettes of the same design and similarly conditioned. If the cigarette weight is within specifications, a mark is made on the cigarette's paper seam 15 mm from the tobacco end. The mark is made with a #2 or softer graphite pencil. The cigarette filter is inserted into the vacuum draw apparatus rubber diaphragm and held in a horizontal position. Start the vacuum draw apparatus, and make sure the center of the rotameter's indicator ball is within ± 50 ml/min of 1000 ml/min. Immediately ignite the cigarette with a preadjusted butane lighter.

    The ignition flame is held to the tobacco end of the cigarette for three seconds to achieve uniform ignition. The lit cigarette is then removed from the diaphragm, held vertically, coal up, under a 600 ml beaker and moved to the test chamber. With the chimney on the test chamber covered, the door is opened, and the lit cigarette is placed vertically, filter down, into a holder located on the center of the substrate assembly. Simultaneously, the door is closed gently and the chimney cover removed. Smoke from the cigarette should pass directly out of the chamber stack. If the cigarette should self-extinguish while in the cigarette holder and before removal to be placed onto the test substrate, terminate the test and identify the results as a self-extinguishment. When the cigarette has burned to the 15 mm mark, simultaneously cover the chimney and carefully open the chamber door. With care, the cigarette is removed from the holder, and the holder is placed in a front corner of the test chamber. The cigarette, <u>with ash still attached,</u> is gently placed onto the top of the substrate assembly so that the coal is located at the geometric center of the surface and the cigarette angled 45° to the fabric warp (*i.e.*, diagonally). See Figure B-3. The cigarette paper seam shall be turned up. Do not drop the cigarette onto the substrate, and do not press the coal onto the substrate. Without delay, simultaneously remove the chimney cover, and gently close the door. A stopwatch is started. If the ash falls off during any part of the transport or positioning process, immediately extinguish the cigarette and begin with paragraph 7 in this section.

8. The burning cigarette and substrate are observed. The smoke plume near the cigarette should remain undisturbed; if not, note your observations on the test sheet. A time record is kept from when the cigarette is placed onto the substrate assembly until either:

   (1) self extinguishment (the coal goes out before the tobacco column is consumed),

   (2) non-ignition (the tobacco column burns to the end without causing the substrate to smolder),

   (3) borderline ignition (ignition appears, but a char does not propagate more than 10 mm, then goes out, while the tobacco column burns to completion), or

   (4) ignition (the substrate starts to smolder, and the char propagates away from the burning tobacco column by at least 10 mm).

   The first three of these observations are considered to be non-ignitions. The fourth is the only outcome that is considered to be an ignition.

   It may be helpful to darken the room intermittently to observe if ignition is occurring. Record ignition time on the test sheet. After recording the ignition time, observe the substrate for 2 to 3 minutes to insure that smoldering has begun. Observations are also made to document the area of fabric char expansion to at least a 10 mm point from where ignition occurred.

9. Smoldering substrates are extinguished with water from a squirt bottle or by prodding the smoldering area with a candle. The time at which extinguishment occurs is recorded. It should be noted that candles are generally used when specimens are to be retained for further study. If candles are used for extinguishing, be sure that all combustion processes have stopped before storing the test assembly. All substrates used in testing are stored in metal containers until they are disposed of in a safe and acceptable manner as determined by the testing facility.

10. After a test is completed, leave the test chamber door open to allow air to circulate throughout its volume. After the chamber has been cleared, preparations may begin for additional testing.

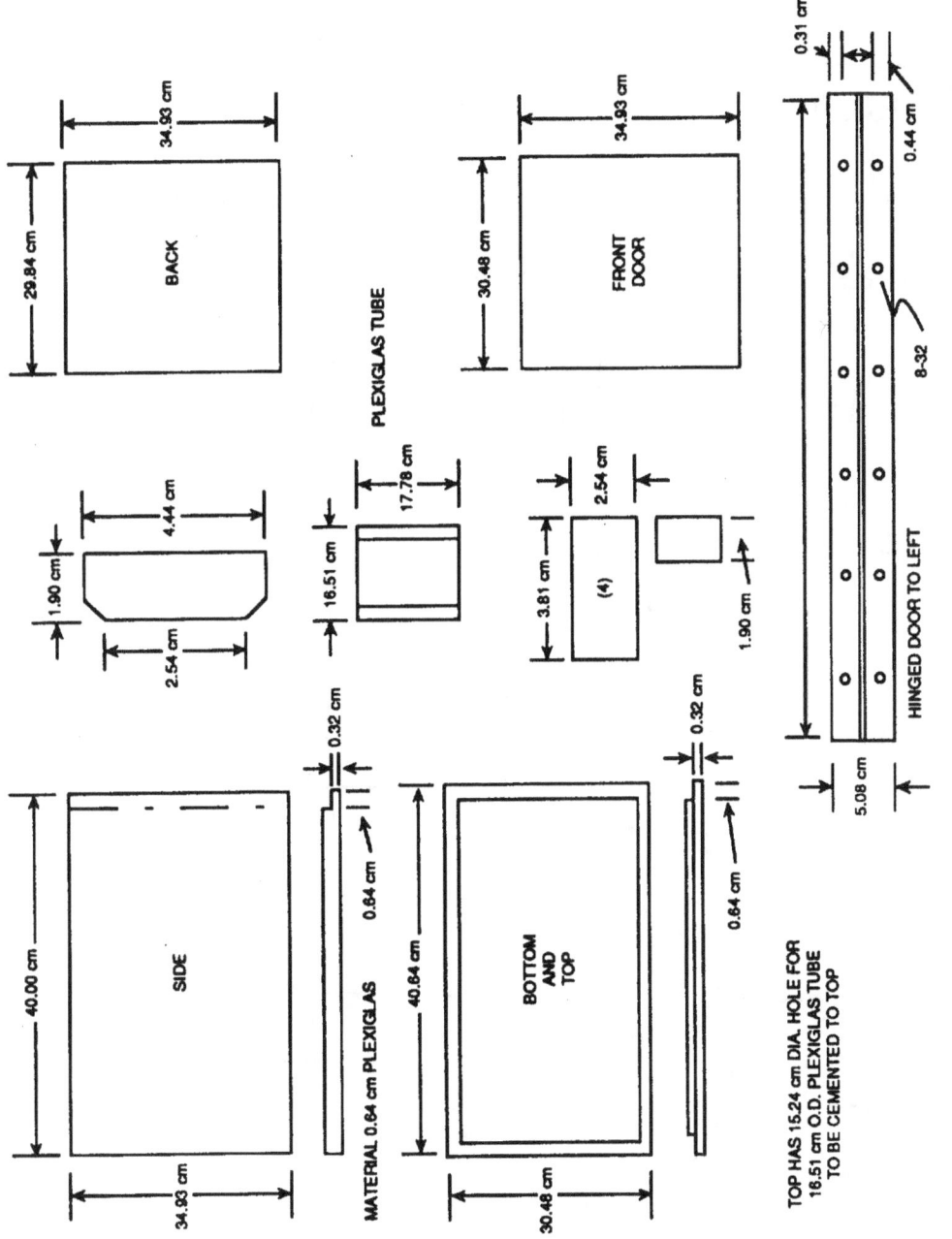

Figure B-1. Schematic of Test Chamber Components.

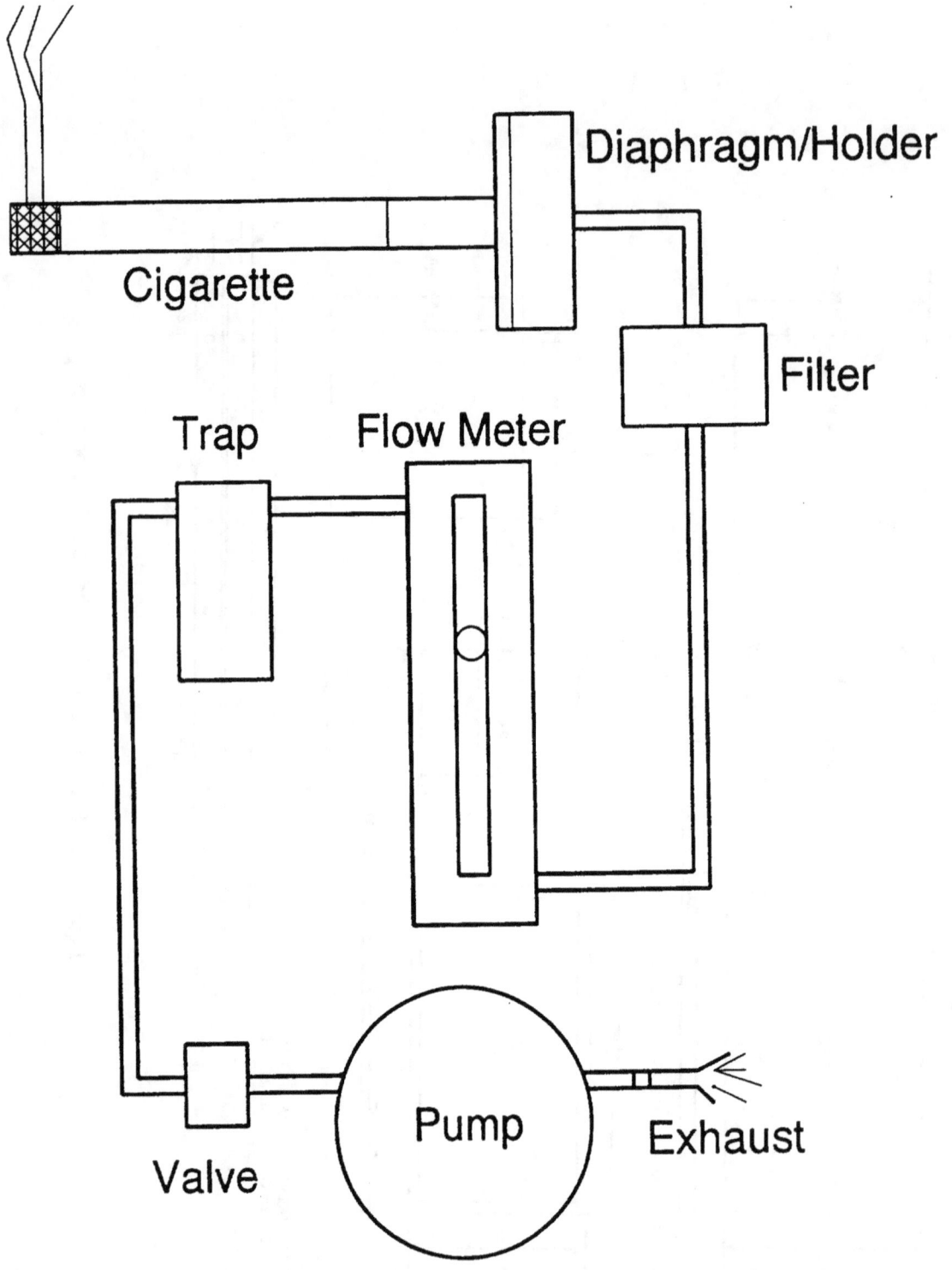

Figure B-2. Schematic of Vacuum Draw Apparatus.

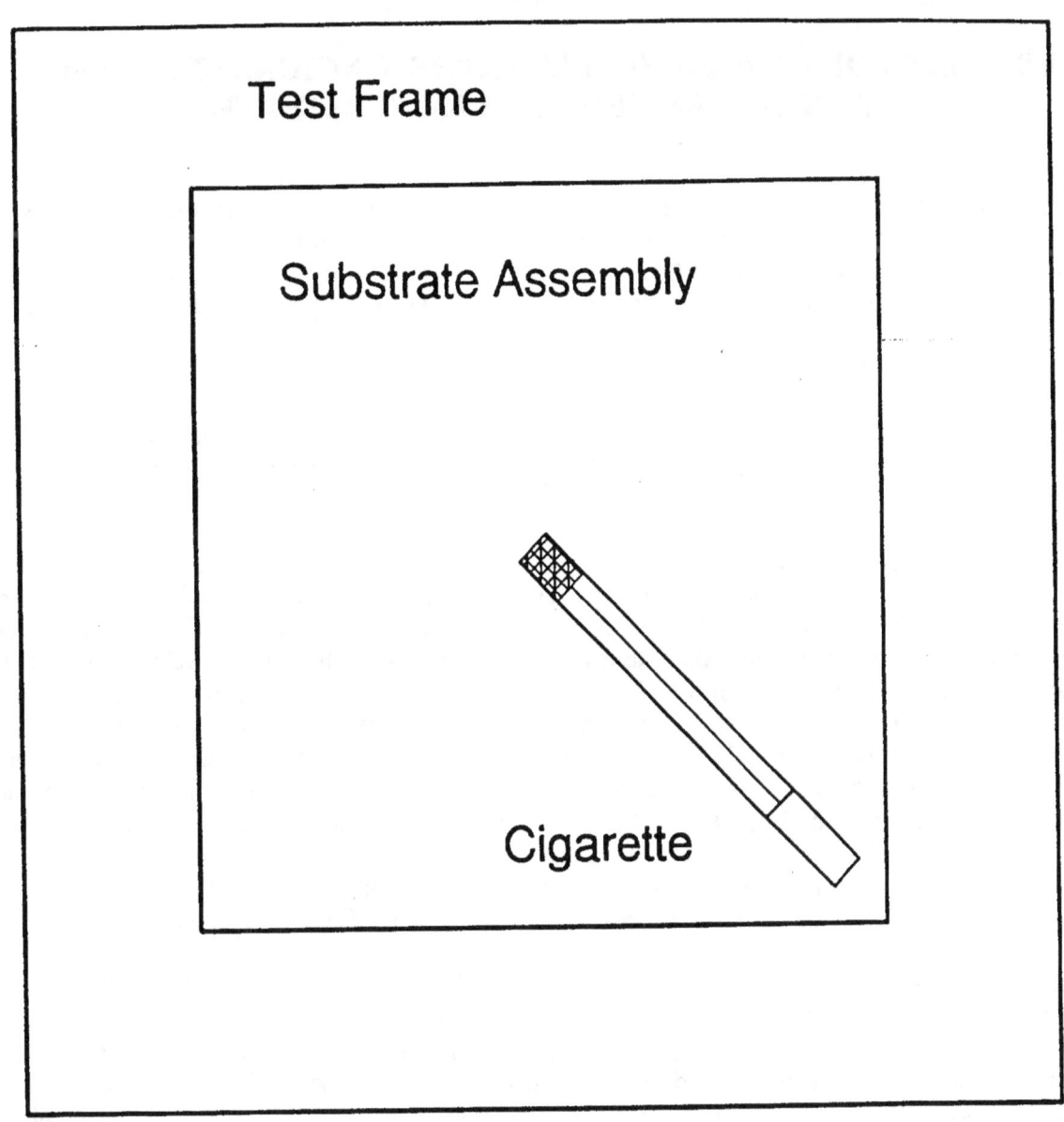

Figure B-3. Location of Cigarette on Mock-Up Method Substrate Assembly.

B-9

APPENDIX C

ESTIMATE OF OXYGEN SUPPLY PATHS TO CIGARETTE COAL ATOP A FLAT UPHOLSTERY SUBSTRATE

The physical system considered here is a lighted cigarette smoldering atop a flat mock-up consisting of a fabric over a polyurethane foam. The issue of concern is the relative importance of oxygen supply to the cigarette coal via two paths: (1) from ambient air above the fabric plane and (2) from ambient air contained within the polyurethane foam below the fabric plane. If the coal receives an appreciable portion of its oxygen supply from below the fabric plane then the ability of oxygen to penetrate the fabric will be important to the behavior of that coal.

This system is three-dimensional and time-dependent; this precludes any exact approach to assessment of the oxygen supply pathways of interest. It will be necessary to make estimates based on the information available.

The total oxygen demand of the smoldering coal is easily calculated from its mass burning rate and the overall stoichiometry of the smolder process. The mass burning rate of the TSG cigarettes in free burn varied in the range $5.7 - 15.5 \times 10^{-4}$ g/s [C-1]. A cigarette studied by Baker [C-2] appears to fall close to the middle of this range (taken here as 10×10^{-4} g/s), using the smolder velocity he reports and an estimated packing density of 0.3 g/cm^3. The TSG study showed that the smolder velocity of a cigarette placed on a horizontal substrate of the type considered here will slow down about 20-25%. Thus the oxygen demand is slowed proportionately. For Baker's cigarette we thus estimate the mass burning rate on the substrate to be approximately $7.5 - 8.0 \times 10^{-4}$ g/s. The stoichiometry of free burn cigarettes has been measured by Muramatsu for a variety of tobacco blends; the results fall in the range $1.4 - 1.9 \times 10^{-2}$ moles of oxygen per gram of tobacco consumed [C-3]; this converts to a range of 0.45 - 0.61 grams oxygen/ gram of tobacco. Then the oxygen demand for the coal smoldering on the substrate is estimated to be in the range $3.4 - 4.9 \times 10^{-4}$ grams of oxygen/second.

Baker measured the oxygen concentration profiles around his cigarette coal in free burn [C-2]. These profiles can be used to estimate the amount of oxygen diffusing from the air above the fabric plane to the coal. These profiles will be changed by the proximity of the substrate; to a first approximation all oxygen which Baker's profiles would show reaching the coal from below (in free burn) is blocked by the presence of the substrate. In addition, the profiles are such that the gradient of oxygen above the coal is substantially less than elsewhere because of the rising smoke plume. Thus, the significant diffusive routes through the oxygen boundary layer on the coal are from the sides and from the front. Inspection of the oxygen profiles presented in reference F-2 suggests that an average oxygen diffusion length on the sides of the coal is 2.8 mm for a distance from 4 mm behind the paper burn line to 4 mm beyond it. Then the mass of oxygen diffusing through these boundary layer profiles is calculated as follows.

$$m_{SD} = A_{SD} \rho D (\Delta Y_{OX} / \Delta x)$$

where m_{SD} is the mass of oxygen entering the coal by diffusion from the sides; A_{SD} is the cigarette surface area through which this oxygen enters (1/4 of the periphery and an 8 mm length yields an area of 1.0 cm^2); ρ is the gas density in the boundary layer; D is the diffusivity of oxygen in the boundary layer; ΔY is the oxygen mass fraction change across the boundary layer; Δx is the boundary layer thickness. Using a mean boundary layer temperature of 150 °C, one makes the following estimate of m_{SD}:

$$m_{SD} \approx (1.0)(7.9 \times 10^{-4})(0.41)(0.2/0.28) = 2.3 \times 10^{-4} \text{ g } O_2/\text{sec}$$

The oxygen entering the front of the coal is more difficult to estimate from Baker's profiles; here it is taken to be comparable to the rate above but modified by the lesser area for diffusion from that direction (0.5 cm^2, the cross-sectional area of the cigarette). This means that the above value is multiplied by 1.5 to get an estimate of the total oxygen inflow from the ambient air above the fabric plane (designated m_{AFP}):

$$m_{AFP} \approx 3.4 \times 10^{-4} \text{ g } O_2/\text{sec}$$

It should be noted that this number is comparable to the estimate above of the total oxygen demand of the coal.

Consider next the issue of oxygen supply to the coal from below the fabric plane. There is essentially no pressure force to cause air to flow upward from the cells of the polyurethane foam, through the fabric and into the cigarette coal. The air within the foam cells is being heated from above; only the finite spatial extent of the heated zone in the lateral (fabric plane) direction implies the application of a weak buoyancy force acting over the very short height of the thermal layer in the foam. This is opposed by the flow resistance of the foam and the fabric. Diffusion of oxygen is less inhibited in the foam because its very open cell structure poses little blockage. Thus oxygen diffusion out of the foam is the main process of interest here. We have no information on the diffusion of oxygen through the less porous structure of a fabric so it will not be included here; the result will be an *overestimate* of the supply of oxygen able to diffuse out of the foam and to the cigarette coal.

The relaxation time of the diffusion process in the foam is short compared to the time required for the coal to move a distance equal to its own length. This means that the oxygen profiles in the foam will be little affected by the fact that the cigarette coal is trying to move slowly over the top of the fabric. Then these profiles can be accurately estimated using the solution to a simple transient problem: diffusion into a sink of finite radius from a semi-infinite medium. This problem is treated in reference C-4; the solution for the spherically-symmetric oxygen concentration profiles as a function of distance from the sink (the cigarette coal) and time (here, time since the coal made contact with the substrate) is as follows:

$$\frac{(C-C_0)}{(C_1-C_0)} = \left(\frac{a}{r}\right) erfc\left(\frac{(r-a)}{2(Dt)^{1/2}}\right)$$

Here C is the mass concentration of oxygen at radius r at time t; C_0 is the initial concentration of oxygen in the foam (essentially same as ambient); C_1 is the concentration maintained from time zero onward at the surface of the oxygen sink (here $C_1 = 0$); a is the radius of the oxygen sink (taken here as 0.35 cm, comparable to the radius of the cigarette); erfc indicates that the error function complement is to be evaluated for the quantity in the brackets.

The oxygen flux of interest is that into the sink (coal) at r = a. This is given by the following:

$$m_{BFP} = -2\pi a^2 D \left(\frac{\partial C}{\partial r}\right)_{r=a}$$

Here the derivative is found from the equation above to yield the following:

$$m_{BFP} = 2\pi a^2 D(C_1 - C_0)\left[\left(\frac{1}{a}\right) + \frac{1}{(\pi D t)^{1/2}}\right]$$

Using a value of oxygen diffusivity of 0.15 cm^2/s in the much cooler and slightly obstructed space filled by the foam, one finds that this reduces to:

$$m_{BFP} = 3.15 \times 10^{-5}\left[2.86 + \frac{1.46}{t^{1/2}}\right]$$

This can then be used to calculate the diffusion flux from the foam space to the coal as a function of time after placement of the cigarette on the substrate. See Table C-1 below.

Table C-1. Calculated Mass Flux of Oxygen from Foam to Coal

Time (sec)	m_{BFP} (g O_2/sec)
5	1.1×10^{-4}
10	1.0×10^{-4}
30	9.8×10^{-5}
100	9.5×10^{-5}
300	9.3×10^{-5}

Since this starts as an infinite diffusive flux at time zero, it is apparent that in only a few seconds it settles down to a nearly constant, low oxygen supply rate. Recall that this does not account for the diffusive resistance of the fabric for which we have no estimate. The actual flux to the coal is thus expected to be smaller than that above, which is already only about 1/4 of the oxygen supply available from above the fabric plane. The implication is that changes in fabric diffusion resistance with changes in fabric structure or weight can be expected to be secondary in their impact on the behavior of the cigarette coal.

Ignition of the fabric itself, below the cigarette coal, probably relies significantly on the above oxygen flux, at least until the reaction zone can spread outward slightly to the point of getting more oxygen from above the fabric plane. Variations in fabric diffusion resistance would not affect this appreciably; the above estimate does not depend on fabric diffusion resistance.

REFERENCES

[C-1] Gann, R., Harris, R., Krasny, J., Levine, R., Mitler, H. and Ohlemiller, T., "The Effect of Cigarette Characteristics on the Ignition of Soft Furnishings," NBS Technical Note 1241, U.S. National Bureau of Standards, Gaithersburg MD, 1987.

[C-2] Baker, R., *Beitr. z. Tabakforsch. Int'l*, 11, 1982, p .181.

[C-3] Muramatsu, M., Umemura, S., and Okada, T., *Nippon Kagaku Kaishi*, 10, 1978, p.1441.

[C-4] Crank, J., The Mathematics of Diffusion, Oxford at Clarendon Press, London, (1956) p. 98.

APPENDIX D

METAL ION CONTENT OF FABRICS: TEST METHOD AND RESULTS

The test method for alkali metal and alkaline earth ions in the cotton duck fabrics consisted of an acid extraction process followed by analysis of the resultant solution by ion chromatography.

Extraction:

1. 100 ml of 4.6 mM HNO_3 poured into 125 ml plastic bottles.

2. 2.5 x 2.5 cm fabric samples immersed.

3. Overnight extraction with agitation at ambient temperature.

4. 5.0 or 10.0 ml aliquot diluted to 50 ml in at least 18.3 megohm-cm water for ion chromatography analysis.

Analysis:

1. Method C-207 from the following source:

   Heckenburg, A., Alden, P., Wildman, B., Krol, J., Romano, J. Jackson, P., Jandik, P. and Jones, W., <u>Waters Innovative Methods for Ion Analysis</u>, Millipore Corporation, Manual #22340, 1989

2. Conditions:

   Instrument: Waters ILC-1 Ion/Liquid Chromatograph.

   Eluent: 0.1 mM EDTA/3.0 mM HNO_3.

   Column: IC-Pak C M/D.

   Flow rate: 1.0 ml/min.

   Injection: 100 µL.

   Standard: 1.0 ppm of Na^+, K^+, Mg^{+2}, and Ca^{+2}.

   Detection: Conductivity.

The results for samples from the three cotton ducks used in this study are shown in the following tables.

Table D-1. Cation Content of #4 Cotton Duck

Sample ID	Cation Content (ppm)			
	Na^+	K^+	Mg^{+2}	Ca^{+2}
4-46-10-1	<10	4608	612	698
4-46-40-2	<10	4688	622	713
4-46-72-2	<20	4429	586	663
4-48-10-5	<10	4273	580	681
4-48-44-1	<10	4201	577	680
4-48-78-3	<10	4254	588	689
4-50-1-1	<20	4475	574	660
4-50-30-1	<10	4372	554	555
4-50-75-1	<10	4535	579	652
4-50-112-5	<20	4525	561	562
4-52-1-3	<10	4600	579	595
4-52-36-5	<20	4646	579	616
4-52-68-2	<10	4457	571	586
4-52-72-3	<10	4372	515	501
4-52-106-2	<10	4654	587	575
4-54-1-2	<20	4587	555	545
4-54-1-3	<10	4477	550	591
4-54-31-2	35	4463	562	581
4-54-66-3	21	4544	563	549
4-54-77-3	22	4569	560	581
4-56-35-6	<20	4527	565	565
4-56-66-1	<10	4447	566	583
4-56-90-3	<10	4550	566	545
4-56-121-5	<10	4514	559	563

Table D-2. Cation Content of #6 Cotton Duck

Sample ID	Cation Content (ppm)			
	Na^+	K^+	Mg^{+2}	Ca^{+2}
6-65-4-1	<10	5571	655	766
6-65-30-2	<10	5423	633	733
6-65-60-3	<10	5659	658	755
6-65-90-6	<20	5905	649	743
6-65-121-5	<20	5775	669	745
6-67-1-43	25	5872	675	753
6-67-32-2	<10	6029	651	743
6-67-60-5	29	5846	661	734
6-67-90-4	32	5766	642	692
6-67-121-2	<10	5986	652	712
6-69-36-4	<10	4443	565	557
6-69-61-1	44	4514	579	584
6-69-95-6	<20	4950	596	622
6-69-120-3	22	4383	551	535
6-71-26-62	<10	5797	640	723
6-71-64-4	<10	5813	642	710
6-71-90-2	29	5765	644	685
6-71-131-1	<20	5591	604	640
6-73-34-5	<20	4276	562	640
6-73-60-4	<10	4501	588	662
6-73-91-6	<10	4540	583	649

Table D-3. Cation Content of #10 Cotton Duck

Sample ID	Cation Content (ppm)			
	Na^+	K^+	Mg^{+2}	Ca^{+2}
10-57-4-3	<10	4474	610	716
10-57-34-1	<10	4551	614	704
10-57-65-2	<10	4471	608	736
10-57-91-4	<10	4463	609	705
10-57-121-3	<10	4562	617	717
10-57-152-3	47	4326	593	683
10-57-182-3	<10	4510	618	709
10-57-182-3	<10	4312	593	682
10-57-216-3	49	4384	602	719
10-57-243-4	22	4393	601	710
10-58-33-3	58	4194	573	684
10-58-61-2	28	4172	576	689
10-58-61-2	51	4318	594	715
10-59-2-23	<10	4342	612	718
10-59-32-4	<20	4369	605	695
10-59-61-3	<10	4522	616	739
10-59-90-2	<20	4465	611	707
10-59-123-1	<10	4348	585	661
10-59-149-1	<10	4481	613	694
10-59-180-3	<10	4274	581	690
10-59-216-1	<20	4559	622	698
10-59-241-1	<10	4435	600	680

Table D-3. (cont.) Cation Content of #10 Cotton Duck

Sample ID	Cation Content (ppm)			
	Na^+	K^+	Mg^{+2}	Ca^{+2}
10-61-32-4	57	4064	577	669
10-61-59-3	<10	4190	579	660
10-61-86-1	<10	4225	595	667
10-61-124-3	<10	4274	605	665
10-61-156-2	<10	4365	594	666
10-63-31-43	121	4081	575	678
10-63-63-33	74	4155	542	605
10-63-96-33	95	4110	593	671
10-63-123-4	95	4110	593	671
10-63-161-1	55	4206	584	687

APPENDIX E

CIGARETTE EXTINCTION TEST METHOD PROCEDURE

Scope:

The purpose of this test procedure is to provide a method for measuring the propensity of cigarettes to ignite specified types of substrate assemblies.

Apparatus and Equipment:

1. Test chambers of the design shown in Figure B-1 shall be constructed for testing the cigarette/substrate combinations.

2. A support holder for the substrate-cigarette assembly shall be constructed; see Figure E-1.

3. A vacuum draw apparatus consisting of the following type shall be used for igniting the test cigarettes: See Figure B-2 of Appendix B.

    The vacuum draw apparatus is composed of a cigarette holder, particulate filter, rotameter, trap, pressure gauge, shut-off valve, and a vacuum pump. The cigarette holder (a plastic drying tube) shall have a flexible diaphragm with a hole cut to a size appropriate for holding a cigarette and making a tight seal around the filter end. The holder must be mounted in a secure fastener which will hold the cigarette firmly in a horizontal position for lighting. The particulate filter, composed of a plastic drying tube 150 mm long and 20 mm in diameter filled with glass wool, shall be adequate to remove smoke from the combustion gases to prevent contamination of all down stream assemblies. The filter shall be changed regularly to insure that gas flow is unobstructed and that contamination is not allowed to pass this point. A rotameter capable of measuring 1000 ml/min, air, shall be used for adjusting flow through the vacuum draw system. A 250 ml impinger trap is used in the system to dampen flow variations. A vacuum gauge capable of measuring to at least 800 mm Hg shall be positioned before the pump in order to track apparatus operating pressure. The electric pump shall be capable of producing a vacuum of 500 mm Hg with a flow rate of at least 23 l/min. A Cole-Parmer Air Cadet pump model 7530-40 or equivalent is found to be suitable.

4. An environmental conditioning room or chamber shall be maintained which provides an area adequate for conditioning cigarettes and test substrate materials. This room shall be capable of maintaining a relative humidity of 55 ± 5% and a temperature of 23 ± 3 °C.

5. A constant humidity box of a size to hold two to three boxes of filter paper and more than one 250 ml beaker of conditioned cigarettes is necessary in the test room if humidity and temperature control are difficult.

6. The conditioning room and test rooms shall be monitored by a recording hygrothermograph (Cole-Parmer Model 8368-00 or equivalent). A Vista Scientific Corporation battery-operated psychrometer or equivalent is used to measure relative humidity in the constant humidity box and to calibrate the hygrothermograph.

7. Chemical or canopy hoods are needed for removing combustion products from the test room. Air flows through these hoods shall be at a level which is necessary to remove cigarette and substrate combustion products while not being high enough to influence combustion processes in the test chambers.

8. Clean plastic or rubber gloves shall be provided for test operators when handling test substrate materials and cigarettes.

9. A laboratory balance capable of weighing to 0.001 g with a repeatability of ± 0.003 g is required for weighing cigarette specimens.

10. A butane gas lighter capable of producing a stable luminous flame no longer than 20 mm is required for igniting test cigarettes.

11. An appropriate fire-proof waste container shall be used for disposal of the ignition test materials.

Calibrations:

The following are guidelines for basic calibrations of test equipment used in this standard. Time intervals stated in this method for calibrations are considered to be minimum. Calibrations of equipment shall be carried out at any time when equipment or test conditions indicate that evaluation and recalibration may be necessary.

1. The ignition test chamber shall be checked before use to insure that the front door seals properly and that air movement in the test area does not introduce transient air movement in the test chamber. Door seals are checked visually to insure that they are closed flush against the chamber's side wall and the latching device secures the door tightly. All construction seams shall be inspected to insure that they are air tight and no cracks shall be visible on any surface of the test chamber. If leaks are detected, measures shall be taken to insure that these areas are again made air tight.

   Stability of air inside of the test chamber shall be determined by mounting a single filter paper in the substrate assembly holder and placing it together with a lighted cigarette in the test position within the test chamber. Observe air movement in the chamber to insure that smoke being emitted by the cigarette is rising vertically and is not showing turbulence within 150 mm above the lit end of the cigarette. If turbulence is noted: 1) the test chamber shall be checked for leaks; 2) the test chamber location shall be evaluated for excess air flow in the laboratory; 3) air flow of the exhaust system shall be evaluated as the source of disturbance. Air flow in the test chamber shall be maintained to produce a near laminar vertical smoke stream to a minimum distance of 150 mm above the cigarette.

2. The vacuum draw apparatus for igniting cigarettes shall be calibrated each week before the beginning of testing. A soap film bubble flowmeter shall be connected to the inlet of the vacuum apparatus at the cigarette attachment point. The vacuum pump is started and adjusted to provide an air flow of 1000 ± 50 ml/min. This flow and range of variation shall be noted and recorded in the laboratory's calibration records for the vacuum system's rotameter.

3. The humidity and temperature sensors used to record environmental conditions in the conditioning room/chamber and test room shall be checked for accuracy each week. This shall be accomplished by comparing their readings with a calibrated psychrometer and thermometer. A record of calibration adjustments shall be kept for each hygrothermograph used in the conditioning and test rooms.

4. The laboratory balance shall be calibrated each week. The calibration is checked by leveling the balance and weighing a calibrated 1.00 g mass. If the balance is found not to be within specifications, make adjustments as needed to bring the device into compliance. Record the results in your laboratory's calibration files.

Test Specimens and Substrates:

Cigarette test specimens and the test substrates are sensitive to contamination. At all times when these materials are handled, clean plastic or rubber gloves shall be worn.

1. Cigarette test specimens shall be protected from physical or environmental damage while in storage. It is important that the specimens not be crushed or deformed in any manner. Measures shall be taken to insure that specimens are not contaminated by foreign materials while in storage, and they shall be protected from degradation by insects. If the specimens are to be stored for more than one week, they shall be placed in a freezer reserved for the sole protection of the cigarette specimens.

2. Substrate materials consist of 150 mm diameter circles of Whatman #2 filter paper. Substrates are formed by placing multiple layers of filter paper into the holder assembly. The filter paper should be placed in the holder assembly with the brass ring on top of the specified number of filter papers to ensure good contact between the layers.

Conditioning:

1. Cigarettes and substrate materials are conditioned at 55 ± 5 % RH and 23 ± 3 °C for at least 24 hours prior to ignition testing. While conditioning, the cigarettes are contained vertically, with filter up, in clean 250 ml polyethylene or glass beakers with a maximum of 20 cigarettes per beaker to assure free air access to the specimens.

2. The substrate filter paper sheets are supplied in boxes of 100 sheets. These boxes shall be opened and placed in the conditioning room along with the cigarettes. There is no need to remove the sheets from the box as long as the top of the box is completely opened. The boxes are positioned to allow air circulation around each box. Each box of filter papers should be conditioned for a minimum of one week prior to use.

3. If the laboratory conditioning room cannot meet the required environmental conditions, a controlled humidity box may be used. Humidity is maintained by the addition of water and chemicals to a holding tray located inside the box. Air in the box must be circulated by means of a small fan in one corner. If a relative humidity measurement is required, the RH and temperature must be measured with a calibrated wet/dry bulb hygrometer. A battery-driven fan instrument such as a Vista Scientific Corp. psychrometer, or equivalent, has been found to be suitable. Humidity measurements are made in the morning and evening of each test day and recorded.

Safety:

1. Exhaust systems should be checked daily to insure that they are working. All products of combustion should be removed from the laboratory work area.

2. Personnel shall be instructed on general emergency procedures in the laboratory and on procedures to handle an uncontrolled fire.

3. Appropriate fire extinguishing equipment shall be provided in each fire test laboratory to extinguish test specimens and to suppress a small fire which may exceed normal controlled limits.

4. An appropriate waste container shall be used in each fire test laboratory for safe disposal of specimens and test assemblies after being exposed to heat and fire.

Test Procedure:

1. Turn on the exhaust system designated for removal of test combustion products 30 min prior to beginning ignition testing.

2. Do not start test preparations until relative humidity and temperature in the test room are stabilized within the following respective ranges 55 ± 5 % RH and 23 ± 3 °C. Record the relative humidity and temperature in your laboratory log.

3. Wear clean, plastic or rubber gloves when handling cigarettes and substrates.

4. Adjust the ignition source, a disposable butane gas lighter, to provide a flame no less than 15 mm in length and no longer than 20 mm.

5. If the relative humidity and temperature in the test room cannot be maintained at the specified test conditions, the filter papers shall be placed into plastic bags and sealed in the conditioning room. They may then be transported to the test room. Test cigarettes may also be transported to the test room in sealed plastic bags.

   If the relative humidity and temperature in the test room cannot be maintained at the specified test conditions, the filter papers shall be placed in the above mentioned humidity box prior to testing.

6. Immediately before selection and ignition of a cigarette for testing, a substrate assembly is removed from its conditioned environment and placed inside the test chamber at the geometric center of its bottom. The brass ring is placed onto the filter paper substrate. Do not use filter papers that will not lay flat in the holder assembly.

7. Without delay, remove a cigarette from the humidity box. Weigh the cigarette and record the results. Discard the specimen if its weight is more than two standard deviations of the mean obtained from weighing 50 randomly selected cigarettes of the same design and similarly conditioned. If the cigarette weight is within specifications, a mark is made on the cigarette's paper seam 15 mm from the tobacco end. The mark is made with a #2 or softer graphite pencil. The cigarette filter is inserted into the vacuum draw apparatus rubber diaphragm and held in a horizontal position. Start the vacuum draw apparatus, and make sure the center of the rotameter's indicator ball is within ± 50 ml/min of 1000 ml/min. Immediately ignite the cigarette with a preadjusted butane lighter.

   The ignition flame is held to the tobacco end of the cigarette for three seconds to achieve uniform ignition. The lit cigarette is then removed from the diaphragm, held vertically, coal up, under a 600 ml beaker and moved to the test chamber. With the chimney on the test chamber covered, the door is opened, and the lit cigarette is placed vertically, filter down, into a holder located on the center of the substrate assembly. Simultaneously, the door is closed gently and the chimney cover removed. Smoke from the cigarette should pass directly out of the chamber stack. If the cigarette should self-extinguish while in the cigarette holder and before removal to be placed onto the test substrate, terminate the test and identify the results as a self-extinguishment. When the cigarette has burned to the 15 mm mark, simultaneously, cover the chimney and carefully open the chamber door. With care, the cigarette is removed from the holder, and the holder is placed in the front corner of the test chamber. The cigarette, <u>with the ash still attached</u> is gently placed onto the top of the substrate assembly so that the filter end of the cigarette is placed between the appropriately sized cigarette anti-roll fingers; see Figure E-2. The cigarette paper seam shall be turned up. Do not drop the cigarette onto the substrate and do not press the coal onto the substrate. Without delay, simultaneously remove the chimney cover and gently close the door. A stopwatch is started. If the ash falls off during any part of the transport or positioning process, immediately extinguish the cigarette and begin with paragraph 7 in this section.

8. The burning cigarette and substrate are observed. The smoke plume near the cigarette should remain undisturbed; if not, note your observations on the test sheet. A time record is kept from when the cigarette is placed onto the substrate assembly until either:

(1) self-extinguishment (the coal goes out before the tobacco column is consumed),

(2) total burn (the tobacco column burns to the end).

It may be helpful to darken the room intermittently or use a dark background behind the test chamber to observe if the cigarette continues to burn. Record the time that the cigarette stops burning on the test sheet. After recording the time, observe the substrate assembly for 1 to 2 minutes to insure that smoldering has stopped. Measurements are made to document the length of unburned cigarette for those not burning to the filter. Record this measurement to the nearest mm.

9. After a test is completed, leave the test chamber door open to allow air to circulate throughout the chamber volume. After the chamber has been cleared, preparations may begin for additional testing.

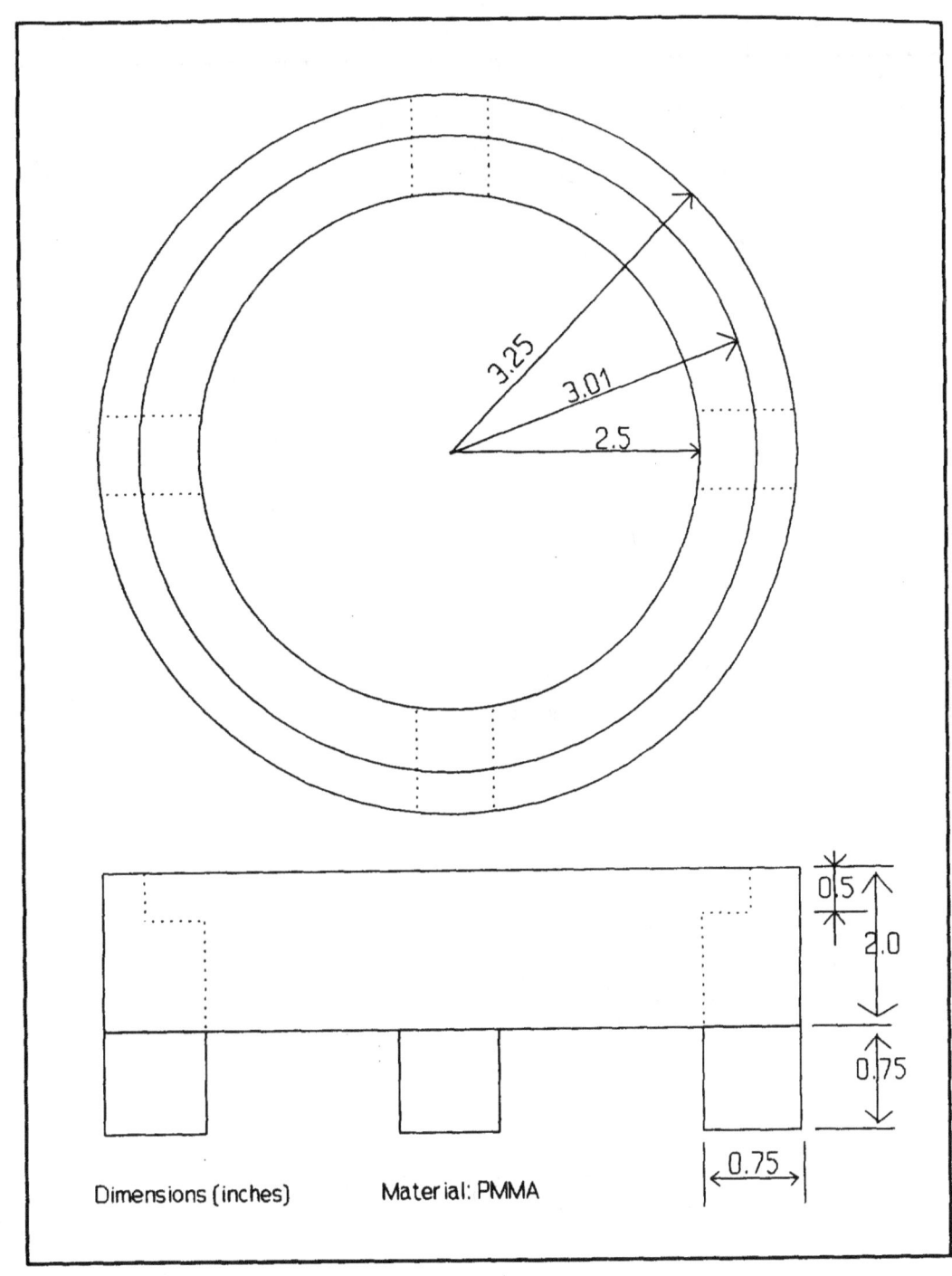

Figure E-1. Details of the Filter Paper Holder Support Structure.

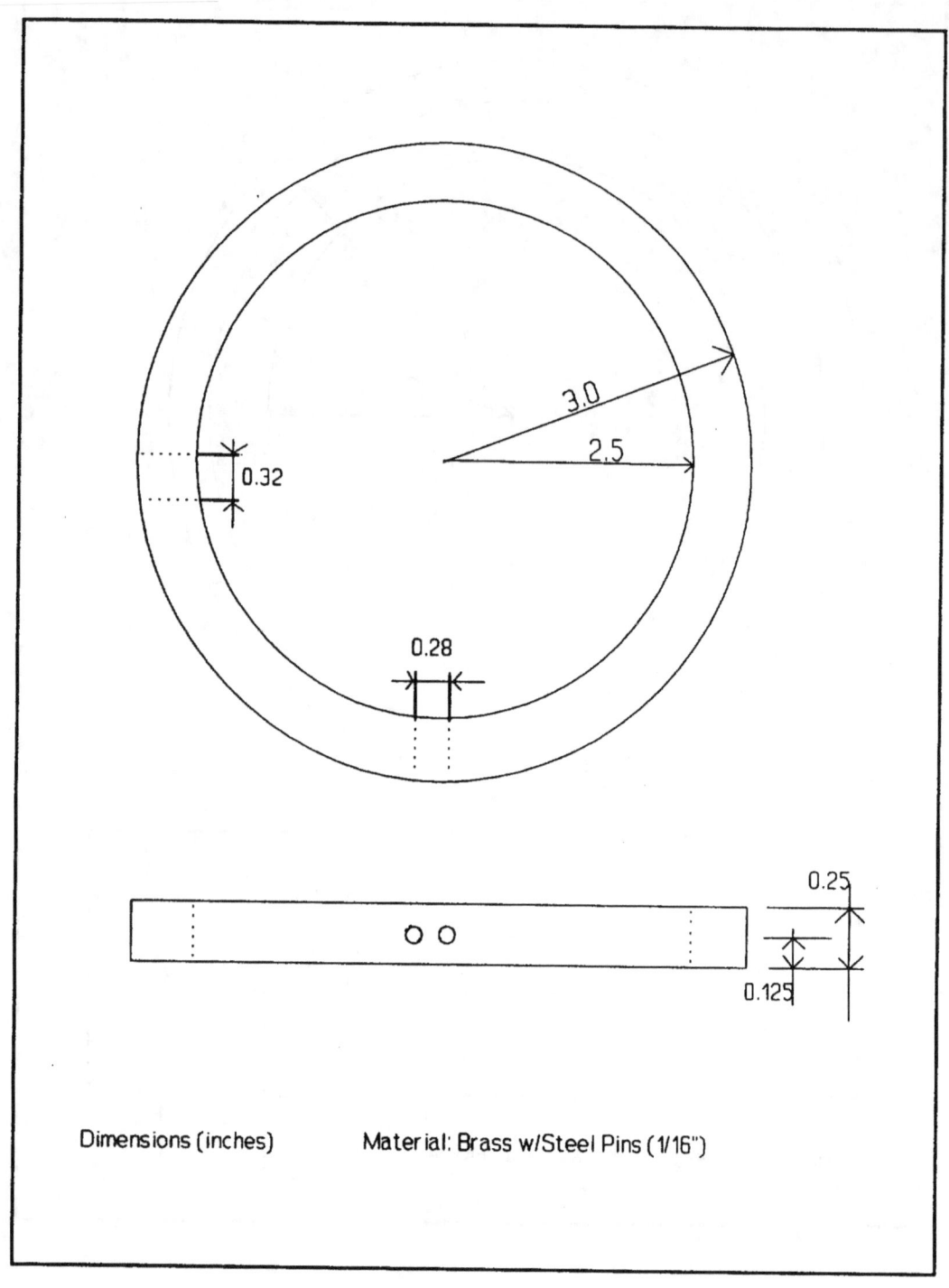

Figure E-2. Brass Holddown Ring and Cigarette Motion Restrainers.

APPENDIX F

REPRESENTATIVE THERMOGRAVIMETRIC DATA
FOR TEST METHOD MATERIALS

Samples of all of the test materials were subjected to thermogravimetric analysis in a Mettler TA 2000C Thermoanalyzer. All tests were run with the samples in aluminum oxide crucibles 8.0 mm OD by 4.5 mm depth. Sample weight in all cases was 2.0 ± 0.1 mg. For the cotton ducks this required four to five pieces of yarn removed from near the location where metal ion concentrations had been measured; these yarn pieces were laid in an open, cross-wise array on the bottom of the crucible. The polyethylene film and filter paper samples were in the form of disks on the bottom of the crucible. The polyurethane foam was in the form of a few small chunks, *ca.* 1 mm, removed with a tweezers from well beneath any previously exposed surface of the foam block being sampled. The heating rate was 5 °C/min. and the atmosphere was 100% nitrogen (< 100 ppm of oxygen). Samples were run up to 80 °C and held there until dry; they were then heated at the noted rate to 500 °C.

The results are shown in Figures F-1 to F-6. In most cases the original data (weight vs. temperature) plus a numerically-derived rate of weight loss are both shown. Samples from two sources of each material are shown. For the polyurethane foam two buns were sampled, as well as two locations within each bun. For the cotton ducks, the potassium level in the adjacent material is reported. Note that the only material showing appreciable variability is the polyethylene film.

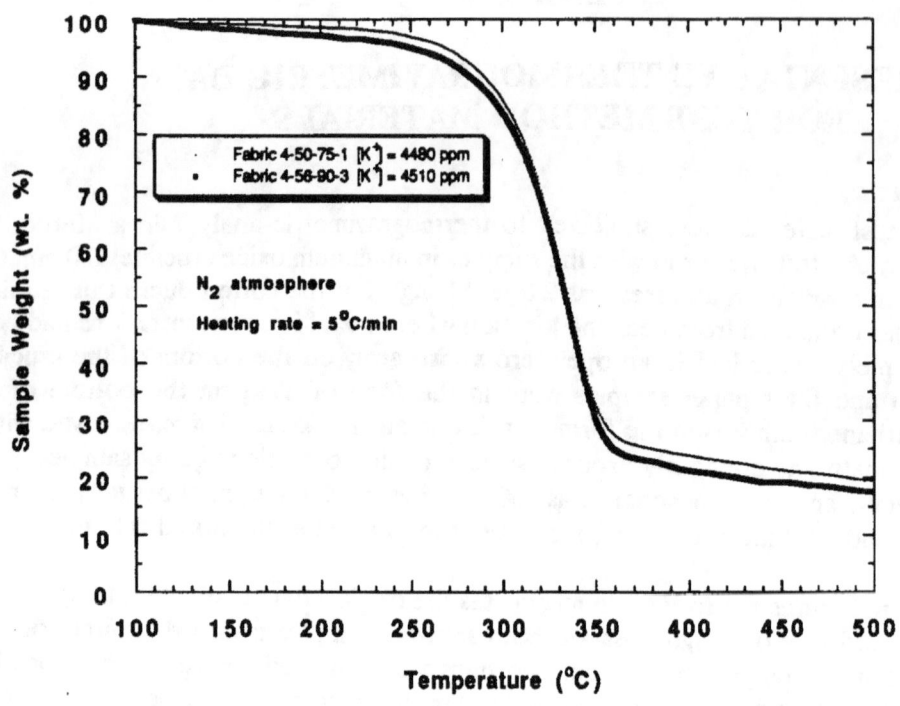
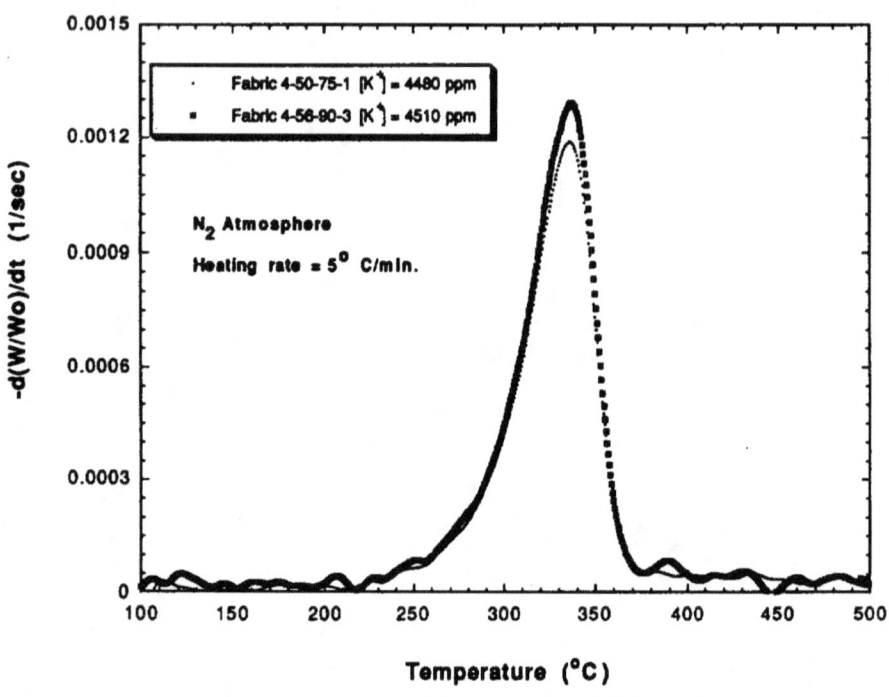

Figure F-1. TG Behavior of Two Samples of Cotton Duck #4.

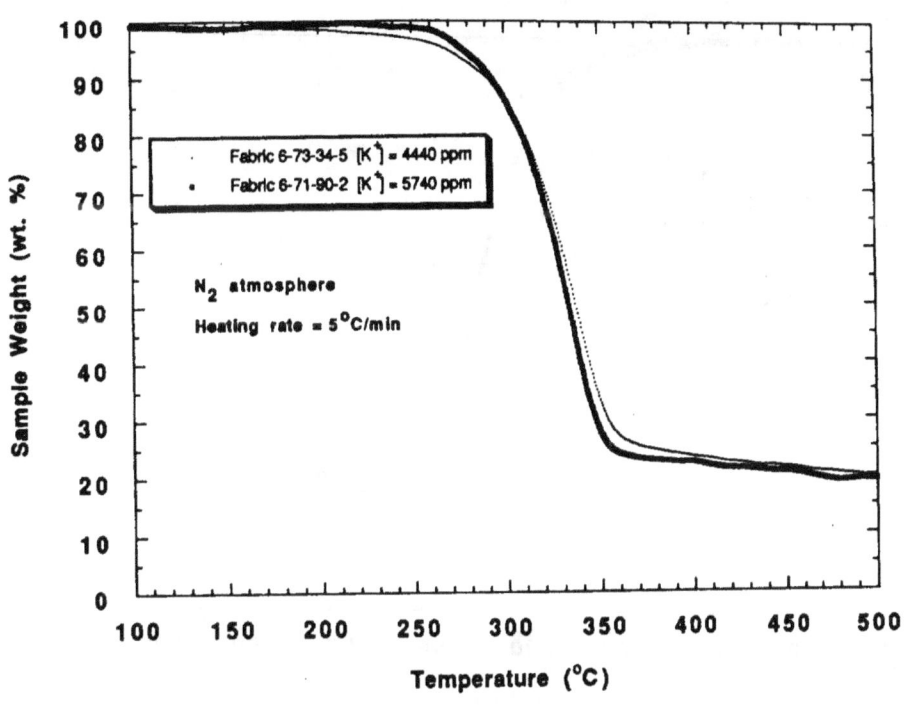

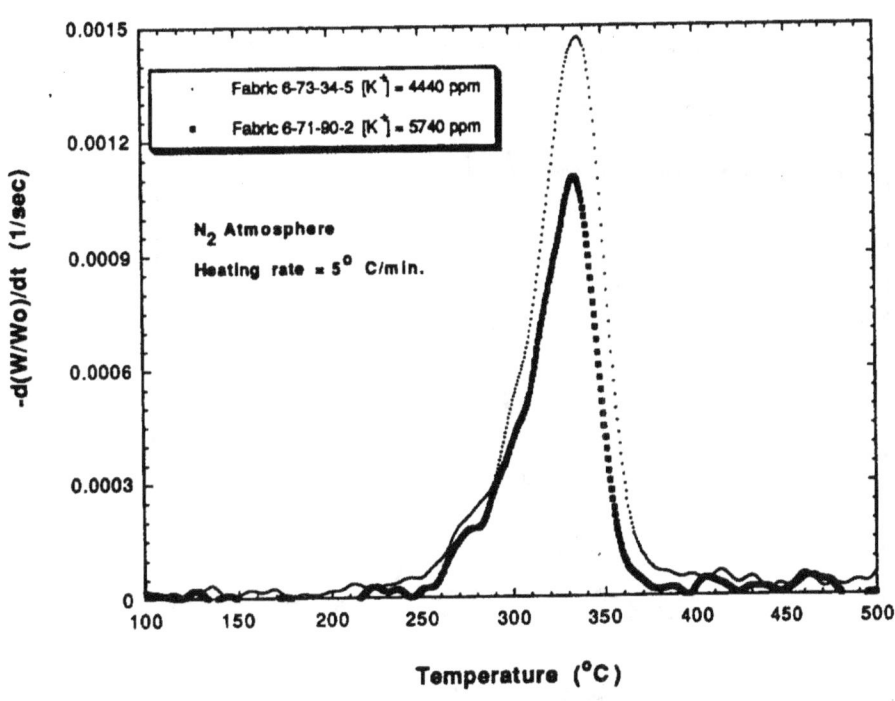

Figure F-2. TG Behavior of Two Samples of Cotton Duck #6.

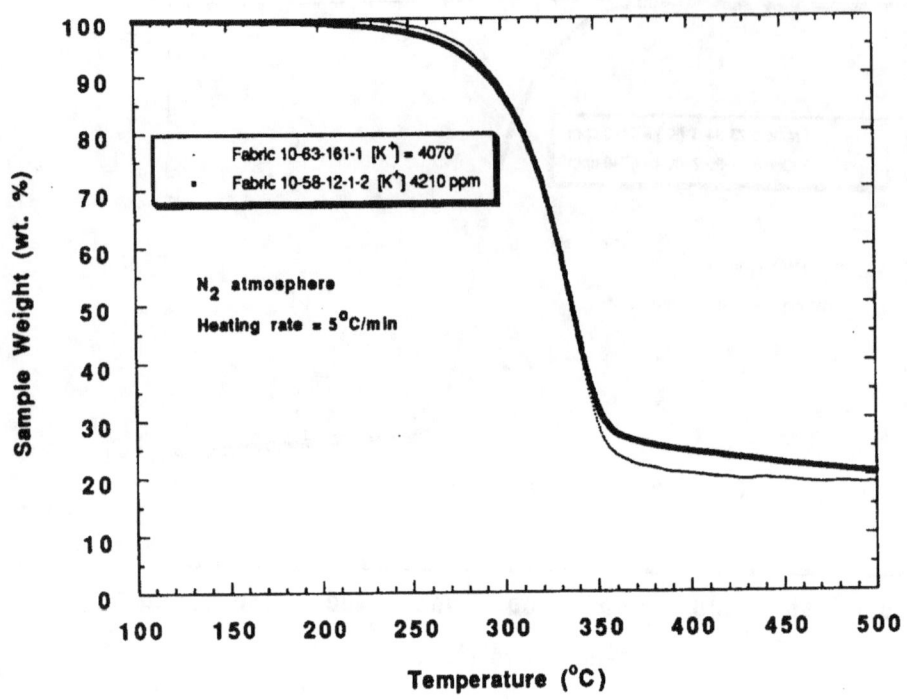
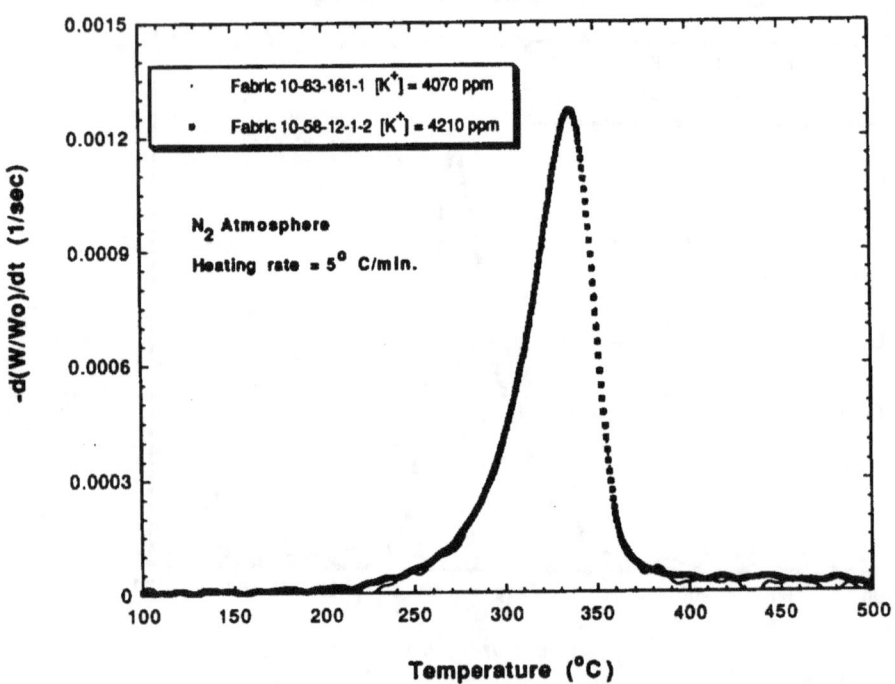

Figure F-3. TG Behavior of Two Samples of Cotton Duck #10.

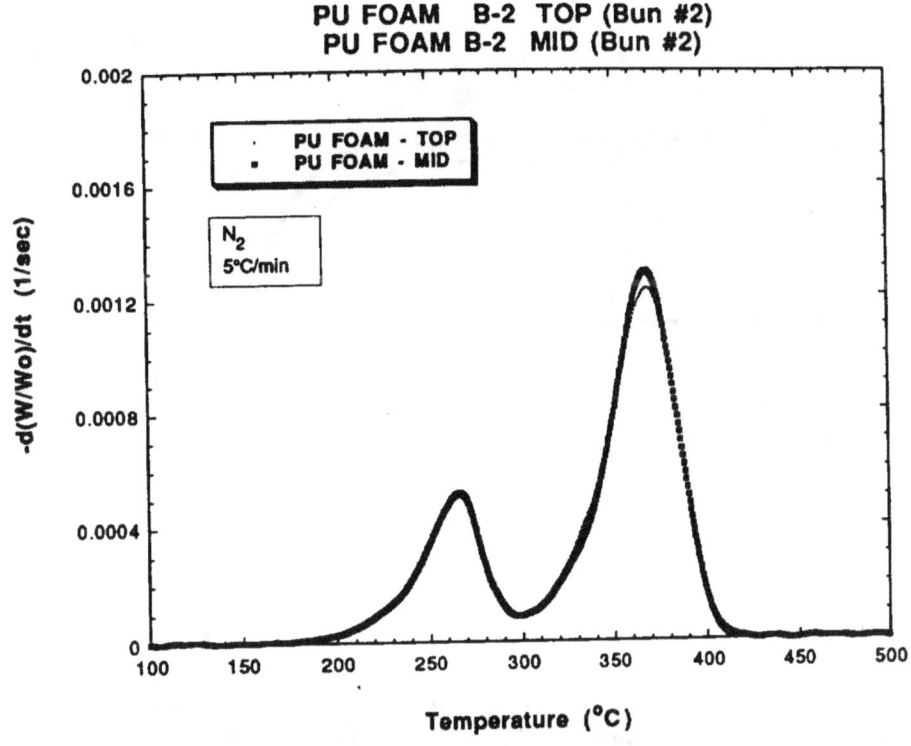
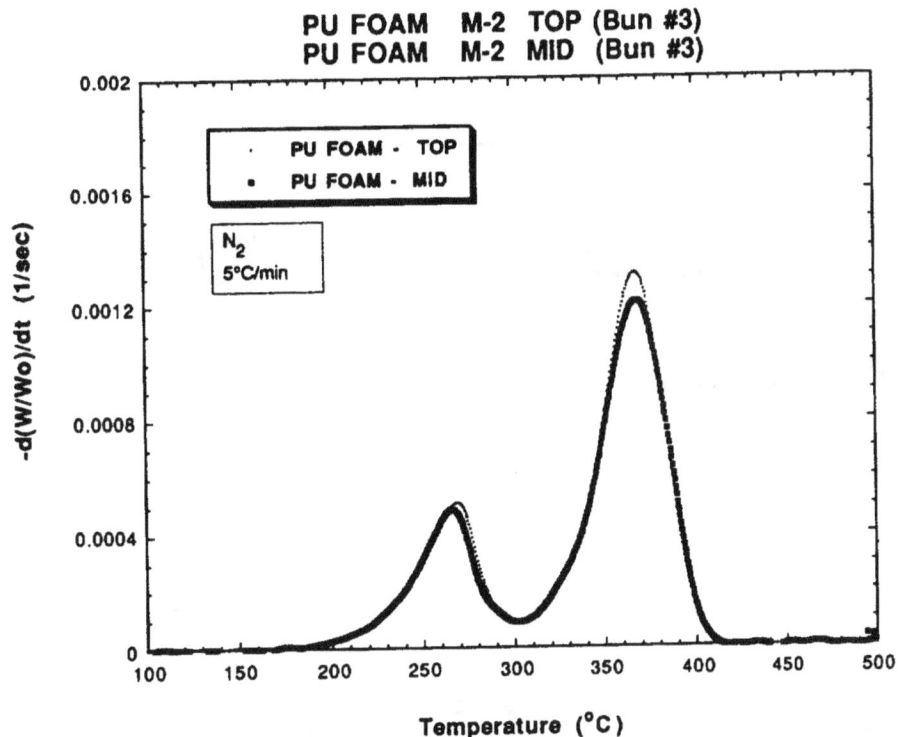

Figure F-4. Derivative TG Data: Two Polyurethane Foam Samples from Top and Bottom of Original Bun.

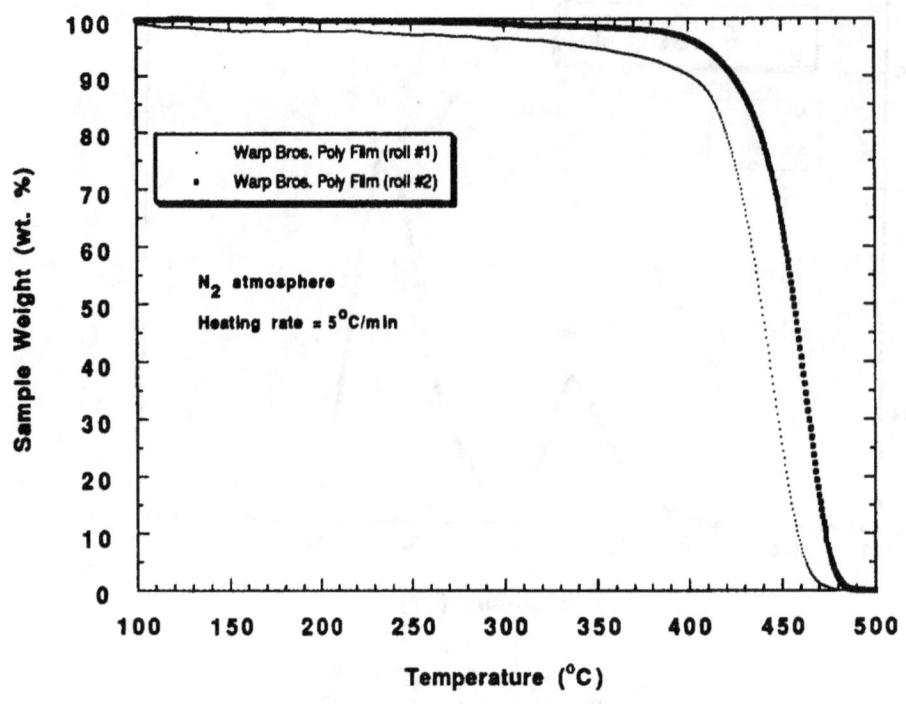
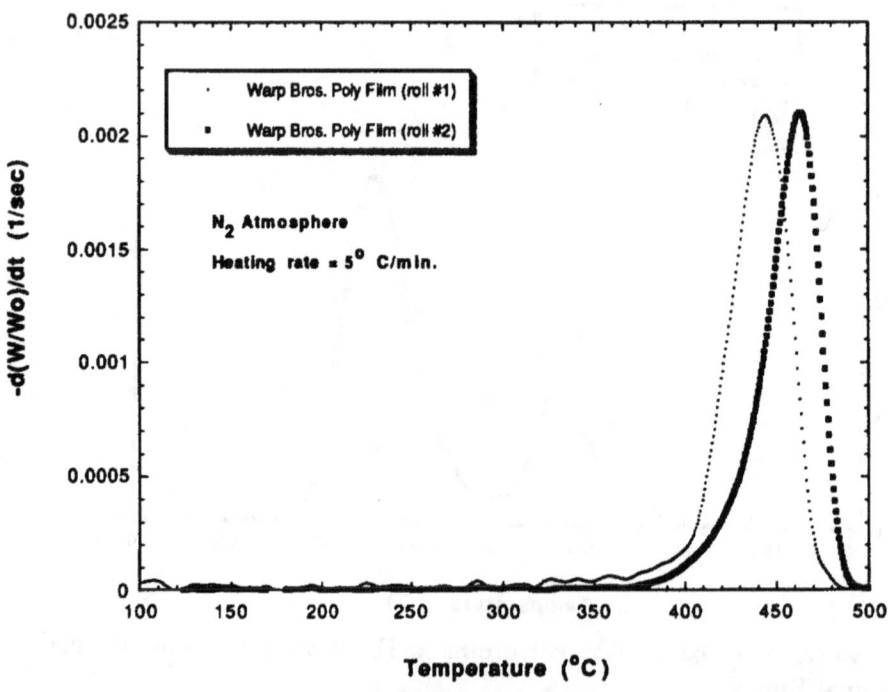

Figure F-5. TG Data for Two Samples of Polyethylene Film.

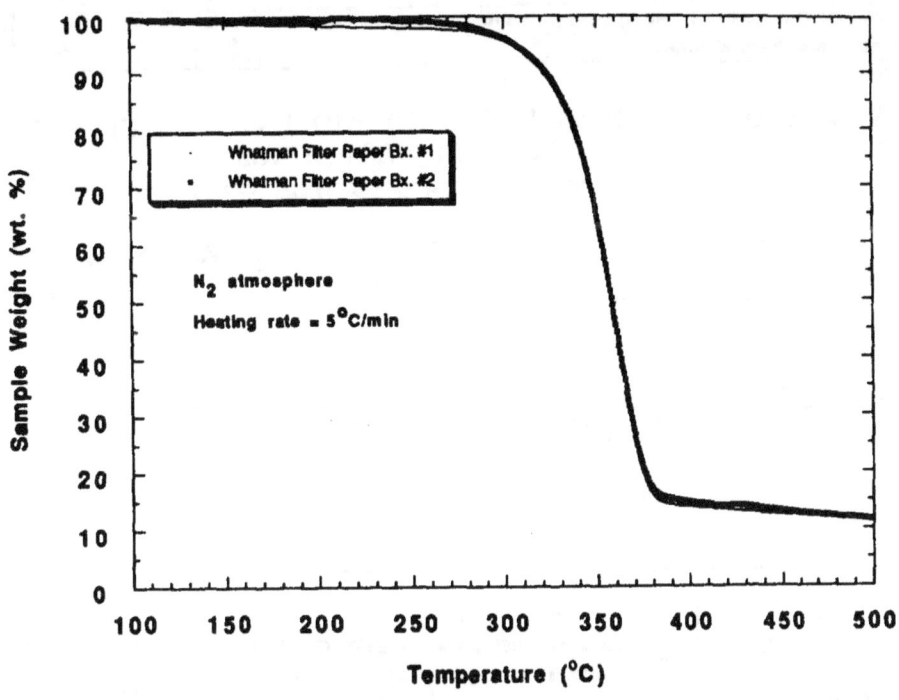
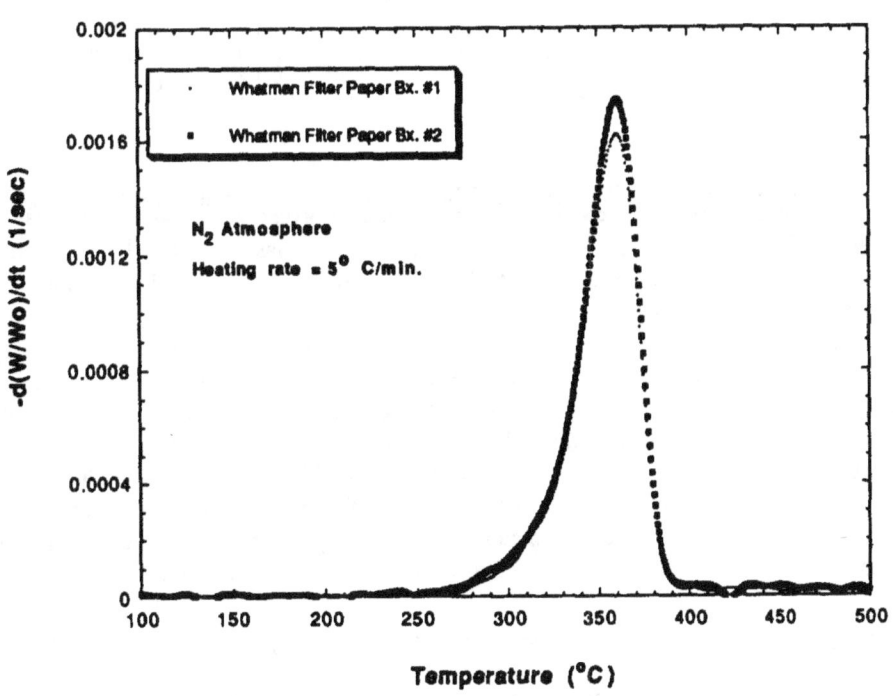

Figure F-6. TG Data for Two Samples of Whatman Filter Paper.

NIST-114 (REV. 9-92) ADMAN 4.09	U.S. DEPARTMENT OF COMMERCE NATIONAL INSTITUTE OF STANDARDS AND TECHNOLOGY	(ERB USE ONLY)	
MANUSCRIPT REVIEW AND APPROVAL		ERB CONTROL NUMBER: W93-1276	DIVISION: 865
INSTRUCTIONS: ATTACH ORIGINAL OF THIS FORM TO ONE (1) COPY OF MANUSCRIPT AND SEND TO: THE SECRETARY, APPROPRIATE EDITORIAL REVIEW BOARD.		PUBLICATION REPORT NUMBER: SP 851	CATEGORY CODE: 240
		PUBLICATION DATE: August 1993	NUMBER PRINTED PAGES: 166

TITLE AND SUBTITLE (CITE IN FULL)

TEST METHODS FOR QUANTIFYING THE PROPENSITY OF CIGARETTES TO IGNITE SOFT FURNISHINGS

CONTRACT OR GRANT NUMBER

TYPE OF REPORT AND/OR PERIOD COVERED

AUTHOR(S) (LAST NAME, FIRST INITIAL, SECOND INITIAL)

Ohlemiller, T. J.; Villa, K. M.; Braun, E.; Eberhardt, K. R.; Harris, Jr., R. H.; Lawson, J. R.; and Gann, R. G.

PERFORMING ORGANIZATION (CHECK (X) ONE BOX)
- [ ] NIST/GAITHERSBURG
- [ ] NIST/BOULDER
- [ ] JILA/BOULDER

LABORATORY AND DIVISION NAMES (FIRST NIST AUTHOR ONLY)

Building and Fire Research Laboratory, Fire Science Division (865)

SPONSORING ORGANIZATION NAME AND COMPLETE ADDRESS (STREET, CITY, STATE, ZIP)

RECOMMENDED FOR NIST PUBLICATION
- [ ] JOURNAL OF RESEARCH (NIST JRES)
- [ ] J. PHYS. & CHEM. REF. DATA (JPCRD)
- [ ] HANDBOOK (NIST HB)
- [X] SPECIAL PUBLICATION (NIST SP)
- [ ] TECHNICAL NOTE (NIST TN)
- [ ] MONOGRAPH (NIST MN)
- [ ] NATL. STD. REF. DATA SERIES (NIST NSRDS)
- [ ] FEDERAL INF. PROCESS. STDS. (NIST FIPS)
- [ ] LIST OF PUBLICATIONS (NIST LP)
- [ ] NIST INTERAGENCY/INTERNAL REPORT (NISTIR)
- [ ] LETTER CIRCULAR
- [ ] BUILDING SCIENCE SERIES
- [ ] PRODUCT STANDARDS
- [XX] OTHER

RECOMMENDED FOR NON-NIST PUBLICATION (CITE FULLY) [ ] U.S. [ ] FOREIGN

PUBLISHING MEDIUM
- [ ] PAPER
- [ ] DISKETTE (SPECIFY)
- [ ] OTHER (SPECIFY)
- [ ] CD-ROM

SUPPLEMENTARY NOTES

ABSTRACT (A 1500-CHARACTER OR LESS FACTUAL SUMMARY OF MOST SIGNIFICANT INFORMATION. IF DOCUMENT INCLUDES A SIGNIFICANT BIBLIOGRAPHY OR LITERATURE SURVEY, CITE IT HERE. SPELL OUT ACRONYMS ON FIRST REFERENCE.) (CONTINUE ON SEPARATE PAGE, IF NECESSARY.)

Research funded under the Fire Safe Cigarette Act of 1990 (P.L. 101-352) has led to the development of two test methods for measuring the ignition propensity of cigarettes. The Mock-Up Ignition Test Method uses substrates physically similar to upholstered furniture and mattresses: a layer of fabric over padding. The measure of cigarette performance is ignition or non-ignition of the substrate. The Cigarette Extinction Test Method replaces the fabric/padding assembly with multiple layers of common filter paper. The measure of performance is full-length burning or self-extinguishment of the cigarette. Routine measurement of the relative ignition propensity of cigarettes is feasible using either of the two methods. Improved cigarette performance under both methods has been linked with reduced real-world ignition behavior; and is reasonable to assume that this, in turn, implies a significant real-world benefit. Both methods have been subjected to interlaboratory study. The resulting reproducibilities were comparable to each other and comparable to those in other fire test methods currently being used to regulate materials which may be involved in unwanted fires. Using the two methods, some current commercial cigarettes are shown to have reduced ignition propensities relative to the current best selling cigarettes.

KEY WORDS (MAXIMUM 9 KEY WORDS; 28 CHARACTERS AND SPACES EACH; ALPHABETICAL ORDER; CAPITALIZE ONLY PROPER NAMES)

Key words: Fire, cigarettes, cigarette test method, ignition, upholstered furniture, statistical analysis

AVAILABILITY
- [X] UNLIMITED
- [ ] FOR OFFICIAL DISTRIBUTION. DO NOT RELEASE TO NTIS.
- [X] ORDER FROM SUPERINTENDENT OF DOCUMENTS, U.S. GPO, WASHINGTON, D.C. 20402
- [X] ORDER FROM NTIS, SPRINGFIELD, VA 22161

NOTE TO AUTHOR(S) IF YOU DO NOT WISH THIS MANUSCRIPT ANNOUNCED BEFORE PUBLICATION, PLEASE CHECK HERE. [ ]

ELECTRONIC FORM